HYDROSTATICS AND MECHANICS

HYDROSTATICS AND MECHANICS

by

A. E. E. McKENZIE, M.A.

Trinity College, Cambridge
Assistant Master at Repton School

"...the moderns... have endeavoured to
subject the phœnomena of nature to the
laws of mathematics..."

SIR ISAAC NEWTON

CAMBRIDGE

AT THE UNIVERSITY PRESS

1934

CAMBRIDGE
UNIVERSITY PRESS

University Printing House, Cambridge CB2 8BS, United Kingdom

Cambridge University Press is part of the University of Cambridge.

It furthers the University's mission by disseminating knowledge in the pursuit of education, learning and research at the highest international levels of excellence.

www.cambridge.org
Information on this title: www.cambridge.org/9781107452572

© Cambridge University Press 1934

First published 1934
First paperback edition 2014

A catalogue record for this publication is available from the British Library

ISBN 978-1-107-45257-2 Paperback

CONTENTS

SECTION II. MECHANICS

PREFACE

This book is the first of three volumes covering the sections into which Physics is usually divided—Mechanics and Hydrostatics; Heat, Light and Sound; and Electricity and Magnetism. The aim is to provide a complete elementary course of Physics from the beginning up to School Certificate and 1st M.B. Standard.

Mechanics is often considered by boys as one of the dullest parts of Physics, despite the fact that it is the basis of many of the greatest engineering achievements. A considerable number of illustrations have therefore been provided and references have been made to the applications of Mechanics, partly to make real the scientific theory and partly to stimulate interest. For "the mind of a child is not a vessel to be filled, but a torch to be kindled".

One of the essentials in science teaching is an adequate supply of suitable problems, which test a boy's mastery of the principles and call for the exercise of thought as well as of memory. Care has been taken in the collection of problems at the end of each chapter and selected School Certificate examples have also been provided.

I am indebted to Mr D. G. A. Dyson for reading the proofs and making many valuable suggestions, for drawing the line diagrams, and also for the ingenuity and time he has expended in perfecting the photograph of the falling and bouncing ball; to my colleague, Mr R. E. Williams, for reading the proofs and generously placing his experience at my disposal; and to Mr W. D. M. Paton and Mr G. S. Dawes for checking the answers to the examples.

Many of the illustrations have been provided by engineering firms and publishers, acknowledgement to whom has been made under the illustrations in question. Mr F. J. Spencer of Stratford-on-Avon kindly allowed me to reproduce the photograph of the pile driver.

Finally I must express my thanks to the following examining bodies who have given me permission to reproduce School Certificate Questions: the Oxford and Cambridge Joint Board, the Northern Universities Joint Matriculation Board, and the Cambridge Local Examinations Syndicate.

A. E. E. M.

Repton
April 1934

SECTION I. MEASUREMENT AND HYDROSTATICS

Chapter I

MEASUREMENT AND UNITS

Introduction. The scientific method.

The aim of this book is to give a simple knowledge of what is meant by Science, and to help those who read it to become in some degree scientists.

It deals with a special branch, known as Physics, upon which the rest of science is based. Bridges, railway engines, motor cars, aeroplanes and wireless sets all depend for their construction and working on the principles of Physics.

The scientific method is to collect as many facts as possible and endeavour to generalise them into a law, or explain them by means of a theory. Usually these facts can only be obtained by means of experiments; indeed, science has been defined as the pursuit of truth by the experimental method.

The detective uses the scientific method in his investigation of crime. He collects all the clues he can, and then invents a theory to fit them. Often his first theory proves incorrect. New facts come to light, showing, perhaps, that his suspicions have fallen upon the wrong person. He then has to invent a new theory. In the same way scientific theories are changed or modified with the discovery of fresh facts.

Many of the greatest discoveries of science have only been possible as a result of very accurate measurements, and the first part of the training of a scientist is the learning of this technique.

Measurement of length.

To measure accurately the length of an object is not so easy as it looks. It is true that to find the approximate length of a bar, all that needs to be done is to lay a ruler alongside, and read the position of both ends. But even in this approximate reading there is a source of error which is liable to give wrong results to those who are not carefully on their guard against it. If the ruler is lying on a table or bench it is essential that the eye should be directly above the end of the bar when its position is being read. If the head is moved it will be found that the reading changes slightly, and this is known as the **error due to parallax.** Fig. 1 shows this. If the ruler is vertical it is likewise necessary

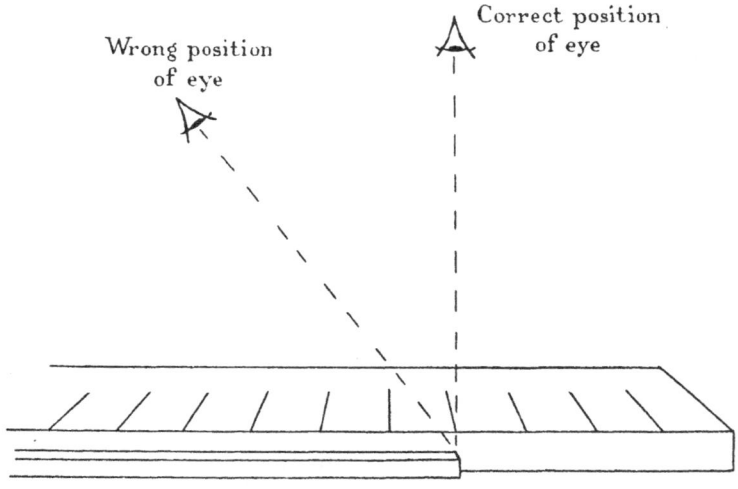

Fig. 1. Error due to parallax.

to have the eye at the same horizontal level as the reading taken, even though this may necessitate climbing on a stool, or kneeling on the ground.

The vernier.

When measuring with a ruler one can attempt to estimate by eye to a tenth of a millimetre, but even after much practice this is difficult and uncertain. By fitting a small sliding scale to the ruler it is possible to measure to this degree of accuracy quite easily. Such an arrangement is called a vernier, after its inventor.

The sliding scale is of length 9 mm., but is divided into ten divisions, so that each division is of length 0·9 mm.

To use the instrument the sliding scale is pushed right up against the object to be measured, as in Fig. 2. The length of the object in the figure is seen to be 3 and a fraction. The fraction is found by looking along the sliding scale and noting which mark is most nearly opposite a mark on the main scale. In this case it is 7, and we take the length of the object as 3·7.

It is a good plan to make a vernier for yourself from two pieces of cardboard, with wide divisions that can easily be marked off accurately. The divisions on the main scale should be 1 cm. long and those on the sliding scale 9 mm. The result will look very

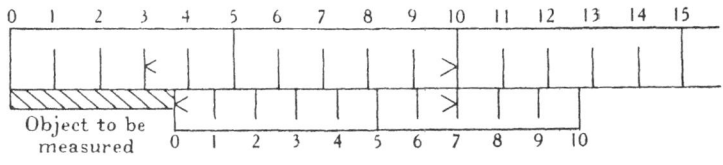

Fig. 2. Vernier.

like Fig. 2. This vernier will read correct to the nearest millimetre and its accuracy can be tested by measuring a body first with the vernier and then with an ordinary ruler.

Fig. 3 shows steel vernier calipers being used to find the diameter of a penny (calipers consist of jaws with which to grip an

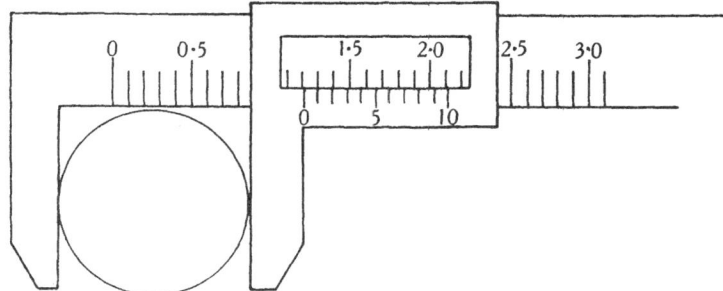

Fig. 3. Vernier calipers being used to find the diameter of a penny (in inches).

object). It is graduated in inches (drawn to a reduced scale). What is the diameter? Read first the position on the main scale of the zero mark of the vernier scale; this will give you the length to the first decimal place. Now look for the mark on the sliding scale which will give you the second decimal place. The answer is given in a note at the end of the chapter. Do not refer to it until you have taken your own reading.

Why the vernier works.

Look at Fig. 2. Why does it follow that the fraction to be measured is 0·7 simply because the 7 mark on the vernier is opposite a mark on the main scale?

Two pairs of arrowheads have been marked in the figure. The distance between the upper pair is 7 and that between the lower pair is $7 \times 0·9 = 6·3$. The difference between these two distances, which is the required fraction, is thus $7 - 6·3 = 0·7$.

The object extends slightly beyond the 3 mark on the main scale, and the sliding scale starts with this extension as a handicap. It gained 0·1 in every division. Since it requires 7 divisions to catch up, its handicap is $7 \times 0·1 = 0·7$. This explains why each division on the sliding scale is made 0·1 shorter than the divisions on the main scale.

Let us consider one other way of looking at this problem. Suppose the sliding scale pushed right home so that the 0 marks on the two scales are opposite. Now move the sliding scale so that the 1 marks are opposite. It has been moved 0·1. When the 2 marks are opposite it has been moved 0·2, and when the 7 marks are opposite it has been moved 0·7, so that the gap between the ends of the two scales is 0·7.

The micrometer screw gauge.

If we wish to find the diameter of a hair or the thickness of paper, even a vernier is useless. We make use of a micrometer screw gauge (see Fig. 4), which can measure to one-hundredth of a millimetre.

Before reading the following description you should get hold of one of these instruments and handle it.

It consists essentially of a very accurate screw with a pitch of, say, $\frac{1}{2}$ mm. The pitch is the distance between the mid-points of consecutive threads, measured parallel to the axis. When the screw is turned through one complete revolution it moves forward or backward a distance equal to the pitch. You can imagine, as a large scale model, a spiral iron staircase which you are descending. When you have gone once round the pillar carrying the staircase, you are standing immediately below the point at which you started, and the vertical distance you have descended is the pitch of the staircase.

Now the head of the micrometer screw is divided into 50 equal divisions, so that each division corresponds to one-fiftieth of a turn, or a forward or backward movement of $\frac{1}{100}$ mm.

When using the instrument the first thing to do is to find the pitch of the screw. Instead of a pitch of $\frac{1}{2}$ mm. it may have a pitch of 1 mm., in which case the head is probably divided into 100 divisions. To find the pitch, turn the screw through four complete turns, and measure with a ruler how far it has moved, whether 2 mm. or 4 mm.

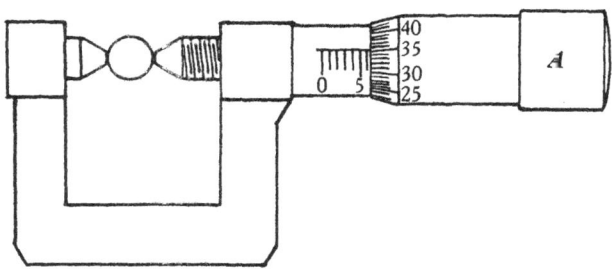

Fig. 4. Micrometer screw gauge.

The "zero error" has now to be determined. Screw the instrument right home, but not too tightly, or you will damage the thread. Usually there is a special cap (A), which, at a certain pressure, turns without moving the screw, and thus prevents damage.

The 0 division on the head of the screw may not be exactly opposite the fixed mark on the frame. This is usually due to a certain amount of play, known as backlash, which has developed through misuse, between the screw and its holder. Note which division is opposite the fixed mark, and determine how many divisions have to be added or subtracted to readings. This is known as the zero correction.

To find the thickness of a piece of cardboard, place it between the jaws of the gauge and screw up. Suppose the gauge has a pitch of $\frac{1}{2}$ mm. Suppose also the reading is 23 divisions, and, in addition, the screw has been turned through one complete revolution. The thickness of the cardboard is 73 divisions, or $\frac{73}{100}$ mm., or 0·73 mm. If the zero correction had been $+1$ division, the true thickness would have been 0·74 mm.

$$
\begin{array}{rll}
50 & \text{divisions} & \text{(one complete turn)} \\
23 & ,, & \text{(scale reading)} \\
1 & ,, & \text{(zero correction)} \\
\hline
74 & ,, &
\end{array}
$$

Thickness of cardboard $=74$,,

Each division $=\frac{1}{50}$ pitch $=\frac{1}{50}\times\frac{1}{2}$ mm.

Thickness of cardboard $=74\times\frac{1}{100}$ mm.

$=0·74$ mm.

Units and standards of length and weight.

Rulers and tape-measures are made in factories, and are copies of standard rulers. But where did the standard rulers come from?

The first standard yard is said to have been the length of Henry I's arm, and the foot, as its name implies, was originally the length of an average human foot. Such vague standards are of no use to an instrument maker, and the first yardstick had to be made by Parliament, and defined as the **British Standard Yard**.

When the Houses of Parliament were burnt down in 1834, the old yardstick was damaged and our modern standard dates from 1844. It is the distance between two marks on a bronze bar at a definite temperature (62° F.). It is now in the custody of the Board of Trade, and exact copies of it are kept in such places as the Houses of Parliament, the headquarters of the Royal Society, the Royal Mint and the Greenwich Observatory.

Other units of length are derived from the yard as follows:

$$
\begin{array}{ll}
12 \text{ inches} = 1 \text{ foot.} \\
3 \text{ feet} \;\;= 1 \text{ yard.} \\
1760 \text{ yards} = 1 \text{ mile.}
\end{array}
$$

The ratios 12, 3 and 1760 are hardly what one would have chosen when inventing the ideal units of length. Chains, rods, poles, perches and the like, while they may have venerable historical traditions, put a further unnecessary tax on one's memory and arithmetic.

There is a much simpler system, the Metric System, which was invented in 1799, just after the French Revolution. The standard unit is the **International Standard Metre**, which is the distance between two marks on a platinum-iridium bar, at the temperature of melting ice. It was originally intended to be $\frac{1}{40,000,000}$ th of the circumference of the earth, drawn through Paris and the poles, but we now know that this is only approximately true.

The other units based on the metre go up in tens. The three important ones are given below:

10 millimetres = 1 centimetre.
100 centimetres = 1 metre.
1000 metres = 1 kilometre.

This system is so simple and logical that it is now universally used by scientists. Its rapid spread over Europe was due largely to the fact that it was imposed by Napoleon's armies. The only other countries in the world besides Great Britain which have not adopted it are the United States and Turkey (where it is optional), China, Paraguay, Canada and the Irish Free State.

At Sèvres, on land given by France as international property, there is a bureau of weights and measures where the standard metre is kept, and also the standard of weight, the **International Kilogram**, which is the weight of a lump of platinum. The International Kilogram was designed to be the weight of 1 litre, i.e. 1000 c.c. of pure water. The corresponding British standard is the **Pound**, the weight of another lump of platinum.

The elaborate precautions to prevent the loss by robbery or fire of the international metre and kilogram read like a story from the Arabian Nights. It must be remembered that the standards are made of platinum, a metal which is hard, and does not rust, but which is so rare that they are worth many times their weight in gold. They are kept in a safe in a cellar, three stories deep, protected by five doors, each with three locks.

Every six years, the international conference on weights and measures meets, and the international standard metre is taken out to compare with the national standards. There are four standard kilograms, one of which is taken out every ten years to compare with the working standard. As handling may affect their weight, the other three are to be disturbed only once every hundred, thousand and ten thousand years respectively.

If ever you feel tempted to handle weights in the laboratory with your fingers instead of the forceps provided, or if through carelessness you drop a weight on the floor, remember the standard kilogram which is not to be touched for 10,000 years.

A new modern standard of length.

Having taken the most scrupulous care to protect their standard of length, it occurred to scientists that despite their efforts the rod might shrink or swell in the course of hundreds or thousands of years, and they devised an apparently permanent and indestructible standard, the wave-length of light.

Light consists of waves in an invisible and weightless medium, called the aether, which fills all space and matter. Imagine a

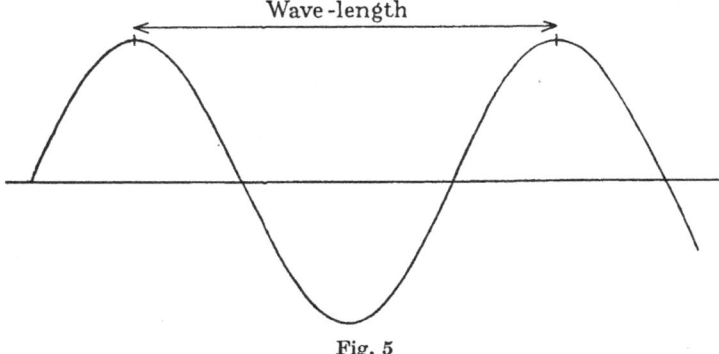

Fig. 5

stone thrown into a pond. From the spot where the stone enters the water there spread out circular ripples which grow in radius and move outwards until they lap the edge of the pond, and are reflected and destroyed. You must imagine the same thing happening in the aether when an electric torch is flashed momentarily in a dark room. But the length of the waves spreading out from the bright filament of the torch is exceedingly minute, just a few ten-thousandths of a millimetre, and they move with almost incredible rapidity, about 186,000 miles per second.

The distance from crest to crest, or from trough to trough, of these aether waves is called the wave-length of light (see Fig. 5).

Professor Michelson, in an historic experiment, counted the number of wave-lengths of light equivalent to the standard metre. The total number was about 2,000,000.

Thus if there were a slight change in length of the standard metre over a long period of years, it could be detected, assuming, of course, that the wave-length of light did not also change. The new standard does not supersede the old, but merely serves as a check.

Measurement of weight.

We can measure weight more accurately than any other quantity. Most laboratories have balances capable of weighing

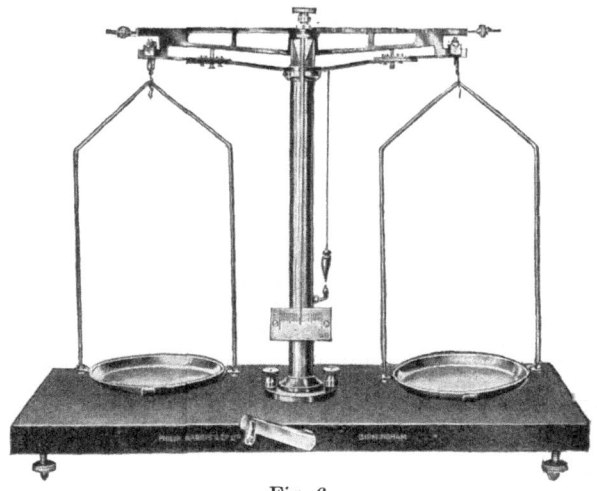

Fig. 6

to 1 part in 10,000. Professor Richards constructed a balance of such extreme precision that he could weigh with it to 1 part in 1,000,000.

Fig. 6 shows the type of balance usually employed. It consists of a beam carrying scale pans at its ends, and balanced at its middle point on a knife edge. This knife edge is usually made of agate, and rests on agate, one of the hardest substances known. In order to save unnecessary wear, it is usually arranged that the beam is not resting on the knife edge when the balance is not in use. Examine a balance carefully, and note how this is done.

To the beam is attached a vertical pointer which swings over a small scale. Before making a weighing, it must be ascertained whether the pointer is swinging about the middle point of this scale. If necessary, adjustments can be made by means of the small screws at the end of the beam or by placing pieces of paper in the lighter pan.

The object to be weighed is now placed in one of the pans, preferably the left-hand pan if the experimenter is right handed and weights placed in the other pan until the pointer once more swings equally on either side of the middle point of the scale. Weights should never be added while the balance is swinging; the beam should be lowered off the knife edge to prevent wear. While noting whether the object is accurately balanced by the weights, the balance should always be swinging. If it comes to rest with the pointer in the middle of the scale, it may be sticking slightly.

The box of weights generally used contains weights from 100 gm. to 1 gm., and smaller flat weights ranging from 500 milligrams, i.e. 0·5 gm., to 10 milligrams, i.e. 0·01 gm. These should always be picked up with the forceps provided; even the cleanest hands are slightly greasy.

With accurate balances it is possible to weigh to 0·001 gm. by means of a small wire rider which is placed on the beam. The right-hand arm of the beam is graduated into ten divisions, and these represent the third decimal place when weighing in grams.

One word of warning is desirable about noting down the weight when the weighing is finished. It is by no means an unknown occurrence for people, after spending ten minutes over a really accurate weighing, to add up the weights wrongly, causing an error of several grams. The following rule therefore is a good one. Add up all the weights in the pan, and record the value in your note book. Then check the addition as you replace the weights in the box.

Summary

The standards of length and weight on the British System are the **Standard Yard** and the **Standard Pound**, respectively. The corresponding standards on the Metric System are the **Standard Metre** and the **Standard Kilogram**.

$$
\begin{aligned}
1 \text{ inch} &= 2\cdot54 \text{ cm.} \\
1 \text{ lb.} &= 453\cdot6 \text{ gm.} \\
1 \text{ kilometre} &= \text{approx. } \tfrac{5}{8} \text{ mile.} \\
1 \text{ kilogram} &= \text{approx. } 2 \text{ lb.}
\end{aligned}
$$

The **vernier** is an instrument for measuring length accurately and the **micrometer screw gauge** an instrument for measuring very small lengths.

Note. Diameter of penny in Fig. 3 = 1·21 in.

QUESTIONS

1. Describe the vernier, how you would use it, and why it works. (A diagram is essential.)

2. What is meant by the pitch of a screw? Explain carefully how a screw is utilised, in the micrometer screw gauge, for measuring very small lengths.

3. What is the standard unit of length on the Metric System? Explain the advantages of the Metric over the British System.

4. A square has sides of length 10 mm. It is measured with a ruler and the possible experimental error is 0·1 mm. The length of the sides may thus be found to be anything between 9·9 mm. and 10·1 mm. Calculate the smallest and largest possible determinations of the area. Hence state the number of significant figures to which one is justified in expressing the area.

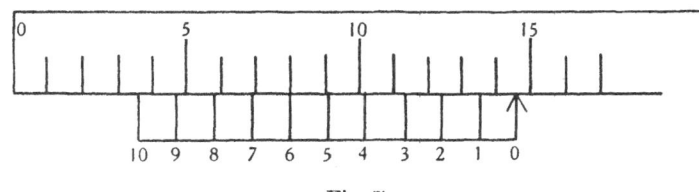

Fig. 7

5. Measure the dimensions of the picture on p. 31 as accurately as you can with a ruler, in centimetres, and find its area. (You should attempt to measure to $\frac{1}{10}$ mm., i.e. to the second place of decimals in cm.) Explain carefully to how many significant figures you are justified in giving your answer.

6. Which is the longer race, 100 yards or 100 metres? There are 30·5 cm. in 1 foot. Express the difference between the races in cm.

7. In a "backward reading" vernier, the sliding scale is graduated from right to left. The position on the main scale of the zero mark on the sliding scale gives the required reading. Thus in Fig. 7 the reading is 14·6. In this vernier the 10 divisions of the sliding scale must have a total length equal to 11 divisions on the main scale. Explain fully why this must be so.

8. A scale for measuring angles is graduated in $\frac{1}{2}°$. A vernier reading to minutes is required. How many divisions will be required on the sliding scale and what must be their total length? ($\frac{1}{2}°$ is equal to 30 minutes.)

9. In the Whitworth measuring machine the pitch of the screw is $\frac{1}{20}$th inch, and the head is divided into 500 divisions. To what distance does each division correspond? If the screw is turned through 1 complete turn plus an extra 279 divisions, through what length has the screw moved?

Chapter II

DENSITY

The three states of matter.

Ice, water and steam are different states of the same substance, and all matter can, in fact, exist in these three forms, **solid, liquid** and **gas**.

Most metals are solid at ordinary temperatures, but one metal, mercury, is a liquid. It looks like liquid silver, and if spilt will run about in drops; in consequence it is sometimes called quicksilver. Lead is quite easily melted over an ordinary fire, but the heat of a blast furnace is necessary to produce liquid iron, and that of the electric furnace to turn it into a gas.

Air as we know it is a gas, but by means of intense cold it can be liquified, and even solidified.

We may distinguish between the three states of matter by saying that solids have a definite shape, liquids have not definite shape, but a definite volume, whereas gases will expand to fill any space, and have not even a definite volume.

Density.

If we obtain small cubes of different metals having the same volume, we find that they all have a different weight (see Fig. 8).

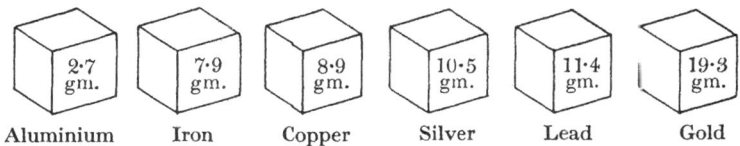

| Aluminium | Iron | Copper | Silver | Lead | Gold |

Fig. 8

Aluminium is one of the lightest of metals and gold one of the heaviest. In order to compare the relative "heaviness" of different substances we use the term **density**, which is defined as **the weight of unit volume.** Thus the numbers in the cubes in Fig. 8 represent the densities of the metals in gm. per c.c.

Other units are sometimes used, such as lb. per cu. ft., when the values will be different. From the definition we can express the density D of a substance in terms of its weight W, and its volume V by means of the simple equation

$$D = \frac{W}{V}.$$

Some advantages of a knowledge of density. Archimedes and Lord Rayleigh.

The densities of many substances, including solids, liquids and gases, have been carefully measured. Whenever a new element is discovered, one of the first things to be determined is its density.

This patient compiling of accurate measurements is a most important feature of science. A book of Physical Constants, containing tables of all kinds, the results of many experiments, is to be found in every laboratory, and is indispensable to the research worker, the inventor and the engineer. The engineer, for instance, can calculate the weight of a bridge from its dimensions, using the density of steel, or materials of which it is composed.

A substance such as a metal can be identified by finding its density and looking up in one of these tables the metal which corresponds to the value obtained. The most famous example of identification by density is the story told of Archimedes, the greatest scientist of ancient times, who lived in Syracuse about 300 B.C. This was in the days before the word "density" had been coined, but actually its conception was used by Archimedes.

The King of Sicily commissioned his goldsmith to make him a crown of a lump of gold. When the crown was made its weight was correct, but the king suspected that the goldsmith had alloyed it with silver, a much cheaper metal, and stolen some of the gold. Archimedes was set to find whether this was so without damaging the crown.

He was unable to solve the problem until one day, on entering his bath, he noticed that the level of the water rose. Whereupon he sprang out, and rushed naked through the streets of Syracuse shouting "Eureka!—I have found it".

He obtained lumps of gold and silver of exactly the same weight as the crown, and placed all three in turn in a vessel brimful of water. In each case he measured the volume of water spilled over, and found that they were all different, thus proving that the crown was neither pure gold nor pure silver. Actually he was demonstrating differences in volumes of equal weights, and thereby differences in density. By measuring accurately the volumes of water displaced he was able to calculate the exact composition of the crown, and inform the king exactly to what

Fig. 9. Archimedes.

extent he had been robbed. (Story taken from Vitruvius, *De Architectura*.)

A more modern and less apocryphal story is the discovery by Lord Rayleigh and Sir William Ramsay of an unknown gas, as a result of very accurate determinations of the density of nitrogen prepared from different sources.

Air is composed mainly of a mixture of nitrogen and oxygen. Lord Rayleigh removed the oxygen, and found the density of the remaining nitrogen. He also found the density of nitrogen prepared from different chemicals. To his surprise he found that these values were not the same; the density of nitrogen obtained from air was 0·0012572 gm. per c.c., while that from other chemicals was 0·0012505 gm. per c.c. This may seem a small discrepancy, but it represents an error of roughly 7 in 1250, that is about 1 in 200 or ½ per cent. Lord Rayleigh was convinced that the errors in his measurements were considerably less than $\frac{1}{2}$ per cent. and in 1892 he wrote a letter to *Nature*, the weekly periodical of British Science, asking whether anyone could suggest an explanation.

A possible cause of the greater density of atmospheric nitrogen was that it might contain a small quantity of a heavier gas which no one before had been able to detect. Lord Rayleigh had made every attempt to remove all impurities from the nitrogen, but he and Sir William Ramsay set to work to find out whether there was one which had been overlooked.

Their experiments resulted in the discovery of argon, an inert gas which is present in all air in very small quantities. Further experiments revealed the existence in air of other previously undiscovered gases in minute quantities, including helium and neon.

These gases, because of their inertness (they will not catch fire, or enter into chemical combination with other substances), have proved very useful. Helium is employed as an alternative to hydrogen in the construction of airships, argon for filling "gas-filled" electric light bulbs, and neon for neon lamps.

Hydrometers.

An instrument which is used for finding the density of a liquid direct is called a hydrometer. The common hydrometer consists of a glass float weighted at the bottom, usually with mercury, having a hollow stem, inside which is a scale. To find the density of a liquid, the hydrometer is floated in it, and the position of the surface level of the liquid on the scale gives its density. Will the top of the scale record a higher or lower density than the bottom?

A determination of the density of the acid is a convenient method of finding whether an accumulator is charged or not. Hydrometers are sold specially for this purpose. Acid is sucked up out of the accumulator, and in this acid a tiny hydrometer or three small coloured balls of different density are made to float. If all three balls float, the density of the acid is high, and the accumulator is charged; if, say, only one floats, and two sink, the accumulator is almost

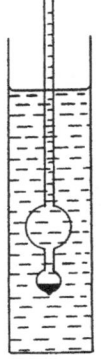

Fig. 10. Common hydrometer.

discharged. Some accumulators used to generate electricity in a wireless set have hydrometers of the latter type fitted into the side, so that one can tell at a glance the state of the charge.

Hydrometers are used by Revenue Officers to find the percentage of alcohol in spirits. The old method was to pour the spirit over a little gunpowder, and endeavour to ignite it. If it did ignite, the spirit was "overproof", and if it did not, "underproof". (Alcohol burns easily, and water does not.) The modern hydrometer method of finding the density of the spirits is far more accurate. Readings have to be taken with great care, since a small error may lead to a considerable difference in duty. The full theory of the hydrometer can only be understood after reading Chapter v. (See Examples 25 and 26, p. 116.)

Specific gravity.

Another term, specific gravity, is sometimes used instead of density. Specific gravity is the number of times a volume of a substance is as heavy as an equal volume of water. Thus if a substance has a specific gravity of 2, any volume of it would weigh twice as much as an equal volume of water.

Definition.

$$\text{Specific gravity} = \frac{\text{Weight of some volume of a substance}}{\text{Weight of an equal volume of water}}.$$

Specific gravity is a ratio; therefore it has no units. It is numerically the same as density, when density is measured in gm. per c.c. Thus, a substance of density 2 gm. per c.c. has a specific gravity of 2. Why is this?

Example. The specific gravity of mercury is 13·6. Find its density (a) in gm. per c.c., (b) in lb. per cu. ft.

(a)　Density of water　　= 1·00 gm. per c.c.
∴　　,,　　,,　　mercury = 13·6 gm. per c.c.
(b)　Density of water　　= 62·3 lb. per cu. ft.
∴　　,,　　,,　　mercury = 62·3 × 13·6
　　　　　　　　　　　= 847 lb. per cu. ft.

Methods of finding density.

Experiment 1. *To find the density of wood or metal in the form of a rectangular block.* It is necessary to find both the weight and the volume of the block. The weight may be found by means of an ordinary balance; and the volume by measuring the length, breadth and depth, and multiplying them together.

Vernier calipers must be used in determining the dimensions of the block if an accurate result is desired.

Experiment 2. To find the density of water. Water having no shape, its volume might be found by pouring it into a hollow rectangular box, and measuring the internal dimensions of the box. It is however more convenient to use a measuring instrument, shown in Fig. 11, called a burette, which is graduated in cubic centimetres. The burette is filled with water, and the clip or tap at the bottom opened to allow water to run through and fill the space below the clip. When measuring the level of the water in the burette it will be found that the surface is curved, and this is known as the meniscus of the water. Always take the reading of the centre of the meniscus, holding a piece of white paper behind the burette. Remember to place the eye on a level with the meniscus to avoid the error due to parallax.

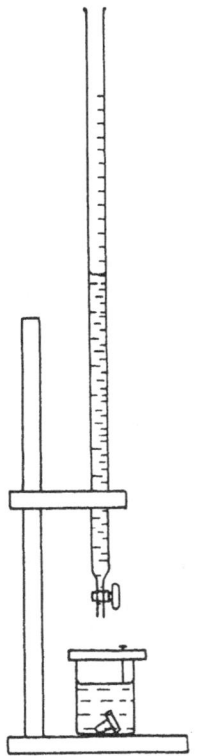

Carefully weigh a glass beaker or can, and run into it 20 c.c. of water from the burette.

Reweigh, and find the weight of the water by subtraction.

If there is time, run out another 20 c.c. into the beaker, and reweigh. Subtract the weight of the beaker and the first 20 c.c. from this weight, and hence find the weight of another 20 c.c. From this you will be able to obtain a second value of the density.

Experiment 3. To find the density of mercury. Use the method of Experiment 2.

Experiment 4. To find the density of an irregular solid, such as a glass stopper. The

Fig. 11

density of a substance can be calculated if its weight and volume are known. The weight of a glass stopper is easily found by means of a balance. The problem is to find its volume. The most suitable

method to adopt in this case is the water-displacement method.
Drop the stopper into a measuring cylinder containing water and
note the rise in level.

A more accurate method is as follows:

Take a small piece of wood through which passes a pin, and
place it on top of a glass tumbler as in the diagram. Fill a
burette with water as in Experiment 2, take the reading, and run
water into the tumbler until its level just touches the tip of the
pin, adding the last few drops with care.

Read the new level of the water in the burette, and obtain by
subtraction the volume of water run out.

Empty the tumbler of water, dry it, and place the glass stopper
inside. Again run water into the tumbler until the level reaches
the tip of the pin, and obtain the volume as before. The difference in the volume of the two lots of water run into the
tumbler with and without the stopper will give the volume of the
stopper.

Hence calculate the density of the stopper.

Experiment 5. To find the density of a liquid. The quickest
method of finding the density of a liquid is to float in it a hydro-

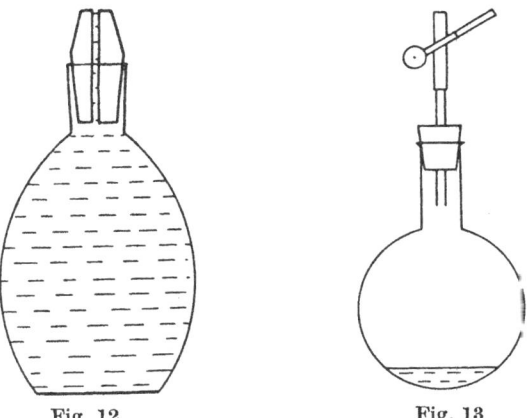

Fig. 12   Fig. 13

meter, and obtain a direct reading. This is not, however, the most
accurate method, and is only possible if there is plenty of liquid.

A special bottle, known as a density bottle (Fig. 12), with a ground glass stopper having a fine hole through it, is made for the purpose. The idea of the stopper is to enable the bottle to be filled with exactly the same volume of liquid every time. This could not be done with a bottle having a neck of ordinary width. Owing to the peculiar skin-like surface of liquids, it might be filled on one occasion with the bottom of the meniscus below the level of the top, and on another with the meniscus above it. (Try this.) On the other hand, if the neck of the bottle were very fine it would be extremely difficult to fill.

Obtain a density bottle, and weigh it empty. Fill with liquid and put in the stopper while it is brimful. (This ensures that there will be no air bubbles inside.) Dry the bottle outside, being careful that the liquid comes up to the top of the hole in the stopper. Reweigh.

Empty the bottle, wash it thoroughly, fill carefully as before with water, dry, and reweigh.

From these three weighings it is possible to find both the weight of liquid and water filling the bottle. Given that the density of water is 1·00 gm. per c.c., see if you can work out the density of the liquid. The bottle is weighed full of water in order to determine its volume.

Experiment 6. *To find the density of air.* It is necessary in this experiment to weigh a vessel full of air, and then to weigh it empty. The obvious method of emptying it of air is to use an exhaust pump, but such a machine is not always available, and there is a way of doing it without a pump.

Take a glass flask, put a little water into it, and fit it with a rubber cork through which passes a short piece of glass tubing (see Fig. 13). Push on to the glass tubing a piece of rubber tubing which can be closed by a clip.

Boil the water vigorously for five minutes or so, taking care before you do so (for your own personal safety) that you have unclipped the rubber tube. The steam will drive out the air, and if immediately after taking away the flame, you close the rubber tube with the clip, you will have your flask empty of air and filled only with a little water and steam.

Cool the flask with cold water under a tap, and incidentally notice if the behaviour of the water is at all peculiar while you are doing this.

Dry the flask very thoroughly, fix it on the pan of a balance (you may have to tie it on with thread), and weigh to the limits of accuracy of which the balance is capable. Some balances have a rider which rests on the right-hand arm, and makes weighing possible to the third decimal place in grams.

Now open the clip, letting in air, and reweigh. By subtracting the first weight from the second, the weight of air filling the flask is obtained.

We have now to find the volume of the air we have weighed.

Pour out the water from the flask into a measuring cylinder and determine its volume. Fill the flask up to the clip with water, and again use the measuring cylinder to find the volume of the water. The difference of these readings will give the volume of the air. The density, or weight in grams of 1 c.c. of air, can now be found.

Experiment 7. To find the density of an insoluble powder such as sand. An interesting exercise in accurate weighing and also in clear thinking is the determination of the density of sand by means of a density bottle.

1. Weigh the bottle empty.
2. Weigh the bottle about one-third full of sand.

Fig. 14

3. Fill up the bottle carefully (see Experiment 4) with water above the sand. Take care that the sand is free from air bubbles. Reweigh.

4. Weigh the bottle full of water only.

Find the weight of the sand. Now try to determine its volume.

Draw the diagrams shown in the figure, and label the bottle, sand and two volumes of water with their correct weights from your readings. Now find the weight of water occupying the same volume as the sand. Given that the density of water is 1·00 gm. per c.c., the volume of the sand follows.

Hence, knowing the weight and volume of the sand, calculate its density.

Write down your readings as in the previous experiments, and explain carefully the steps in the working out of the final result.

SUMMARY

Matter can exist in three states, **solid, liquid** and **gas.**

Density is **weight per unit volume.** It is usually measured in gm. per c.c. or lb. per cu. ft. The density of water is 1·00 gm. per c.c.

Table of Densities (in gms. per c.c.)

Metals

Magnesium	1·74	Silver	10·5
Aluminium	2·70	Lead	11·3
Zinc	7·10	Mercury	13·6
Tin	7·30	Gold	19·3
Iron	7·50–7·9	Platinum	21·4
Copper	8·90	Osmium	22·5

Common solids

Glass (common)	2·4–2·8	Ice	0·92
Wood (yellow pine)	0·4–0·6	Cork	0·22–0·26
(oak)	0·6–0·9		

Liquids

Water	1·00	Turpentine	0·87
Methylated spirit	0·83	Glycerine	1·26
Paraffin	0·80 (approx.)		

Gases (at normal temperature and pressure)

Air	0·0012	Nitrogen	0·00125
Hydrogen	0·000090	Argon	0·00178
Helium	0·00018		

Method of finding density.

1. *Solids.*

(a) Weigh and find dimensions if in the form of a rectangular block.

(b) Weigh and find volume by displacement of water if irregular in shape.

(c) Use a density bottle if in the form of a powder, such as sand.

2. *Liquids.*
   (a) Hydrometer.
   (b) Density bottle.
   (c) Burette.

3. *Gases.*
   Weigh flask empty and full of gas. Find volume by means of water.

QUESTIONS

1. Describe with a diagram how you would find the density of:
   (a) An uncut pencil.
   (b) A quantity of small glass beads.
   (c) Sea water.

2. You are given a small cube of copper and an irregular lump of the same metal which is thought to have a closed cavity inside it. Describe how you would decide, without cutting the metal, whether this is so or not.

3. Define density. The weight of a cork is 10 gm., and its volume is 40 c.c. What is its density?

4. A block of wood measures $5 \times 1 \times \frac{1}{2}$ ft. and weighs 1 cwt. What is its density in lb. per cu. ft.?

5. A gold ring weighs 9·65 gm. If the density of gold is 19·3 gm. per c.c., what is the volume of the ring?

6. The density of glass is 2·6 gm. per c.c. A sheet of glass is square, having a side of length 40 cm., and is 0·25 cm. thick. Find the weight of the sheet.

7. Distinguish between density and specific gravity.
   The densities of fresh and sea water are 62·5 and 64 lb. per cu. ft. (approximately). Find (a) the specific gravity, (b) the density in gm. per c.c., of sea water.

8. Describe how you would find the length of a tangle of wire without unravelling it.

9. A piece of tin-foil is rectangular in shape, being 10 cm. long and 8 cm. wide. If it weighs 29·2 gm. and the density of the tin is 7·3 gm. per c.c., find its thickness.

10. A thread of mercury 12 cm. long weighs 8·16 gm. If the density of mercury is 13·6 gm. per c.c., find the area of cross-section of the thread.

11. A modern type of saucepan is approximately a hollow cuboid without a lid. If the base is a square of side 20 cm., the depth 15 cm. (external measurements), and the metal is 4 mm. thick, find the weight (neglecting the handle), if the saucepan is made of (a) aluminium, (b) cast iron. (Densities of aluminium and iron are 2·70 gm. per c.c. and 7·5 gm. per c.c. respectively.)

12. The sp. gr. of quartz is 2·6 and of gold 19·3. A nugget of gold and quartz is found to weigh 300 gm. and its sp. gr. is 6·5. What weight of gold is there in the nugget?

13. How would you find the density of ice?

14. How would you find in the laboratory the density of coal gas? Give full experimental details.

15. Describe fully how you could test the accuracy of the graduations of a burette, given an accurate balance.

16. 25 c.c. of salt solution of density 1·2 gm. per c.c. are mixed with 35 c.c. of pure water. What is the density of the mixture?

17. 90 gm. of sulphuric acid of sp. gr. 1·8 are mixed with 90 gm. of water. If the sp. gr. of the mixture is 1·5, find the contraction when the liquids are mixed.

18. A density bottle when full contains 40·5 gm. of methylated spirit. If the density of methylated spirit is 0·81 gm. per c.c., find the volume of the bottle.

19. How many grams of glycerine (density 1·26 gm. per c.c.) can be put into a bottle which will hold 100 gm. of sulphuric acid (density 1·8 gm. per c.c.)?

20. The density of an insoluble powder is found by means of a density bottle as in the experiment described on p. 21. Read the account of this experiment, and then work out, by the method suggested, the density of the powder from the following readings:

Weight of empty bottle $= 20$ gm.
Weight of bottle and powder $= 40$ gm.
Weight of bottle and powder and water $= 57$ gm.
Weight of bottle filled with water only $= 45$ gm.

21. A crown composed of a mixture of gold and silver weighs 877 gm. and has a volume of 50 c.c. Find (a) the volume, (b) the weight, of silver in the crown. (Sp. gr. of gold and silver 19·3 and 10·5, respectively.)

Chapter III

PRESSURE IN LIQUIDS

Everyone knows that "Water finds its own level". Suppose that two glass vessels of different cross-section (see Fig. 15) are connected together by a tube at the base, and the whole is nearly full of water. The surface of the water in each vessel is at the same level. But how can it be explained that the comparatively great weight of water in A does not drive the smaller weight of water in B straight out of the tube? Here is the unusual phenomenon of a heavy weight being balanced by a lighter one.

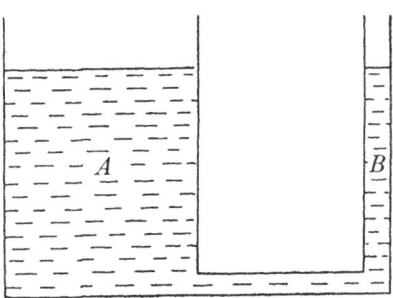

Fig. 15

What is it that is equal in the two columns of liquid? The answer is, the weight of liquid per unit area at any given level. Although the weight of water in A is greater than the weight of water in B, it has a larger area to push down upon. The weight resting on unit area at the bottom of A is equal to the weight resting on unit area at the bottom of B.

Weight, or **force per unit area**, is called **pressure**. The pressure at the bottom of A is said to be equal to the pressure at the bottom of B.

Pressure.

Let us attempt to elucidate the meaning of pressure by means of an illustration.

When a person is standing, his boots press down on the ground with a force equal to his weight. This force is distributed over the area on which his boots are in contact with the ground.

He will be less likely to sink into deep snow if, attached to his boots, he wears skis, which consist of a pair of wooden runners, about 8 ft. long and 4 in. broad. The force he exerts on the snow is now distributed over a much larger area. He is therefore exerting a much smaller pressure on the snow.

Let us calculate the pressure exerted on the snow by a 12 stone man, wearing skis, when standing on one foot. (As soon as he begins to propel himself about, his weight rests alternately on each foot.)

$$\text{Weight of man} = 12 \times 14 = 168 \text{ lb.}$$
$$\text{Area of ski} = 96 \times 4 = 384 \text{ sq. in.}$$
$$\therefore \text{ Pressure on snow} = \frac{\text{Force}}{\text{Area}} = \frac{168}{384}$$
$$= \frac{7}{16} \text{ lb. per sq. in.}$$

Note. Pressure may be measured in lb. per sq. in., lb. per sq. ft., gm. per sq. cm., etc.

Now calculate what pressure he exerts on the snow when walking in boots, having a length of 1 ft., and an average breadth of 3 in.

Compare the pressures exerted by boot and ski, bearing in mind that the forces they exert on the snow are equal, namely the weight of the man.

The difference between force and pressure is also illustrated by the driving of a nail into wood. The sharp point of a nail has a very small area, and consequently when struck by a hammer, the pressure exerted by the point of the nail on the wood is very large. A blunt nail, although struck with equal force, will not penetrate the wood so easily.

Pressure of the domestic water supply.

The characteristic of a liquid is that it can flow. It will only flow, however, if caused to do so by a difference in pressure.

Water flows out of a domestic tap because the level of the water supply is higher than the level of the tap. If the reservoir supplying the water is not at a greater height above sea level than the town or village, machinery is used to pump water from the reservoir to a tank at the top of a water tower, situated where possible at the top of a hill. The pressure which causes the water to emerge from the tap will depend on the difference in height of the tap and the company's water supply.

The pressure in a liquid increases with the depth, and is exerted in all directions.

A liquid, like a solid, exerts a pressure on the vessel in which it is standing, owing to its weight. Unlike a solid, however, a liquid presses on the side walls of its containing vessel.

If a long glass cylinder with two holes at the side, as in Fig. 16, is filled with water, the water is found to issue with greater force from the lower hole than the upper one. **This demonstrates that the pressure in a liquid increases with depth.** There is a greater weight of water forcing the water out of the lower hole than out of the upper, and hence a greater pressure at the lower hole.

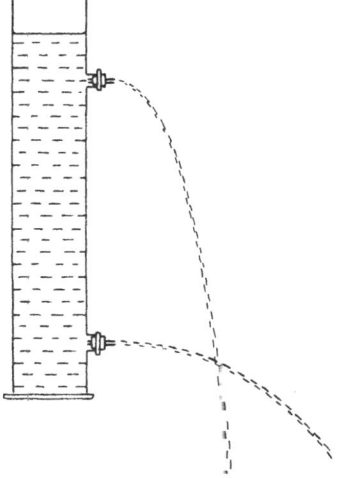

We may learn another fact about liquid pressure from this experiment. The water is being forced out sideways. Thus pressure is transmitted through the water not only downwards, but sideways. Actually, **pressure in a liquid is transmitted equally in all directions.** This can be simply demonstrated by means

Fig. 16. Pressure increases with depth.

of a rubber ball punctured with a number of holes. Fill the ball with water, lay it on a table, and press down upon it. The water squirts out in all directions, although the ball is being pressed in one direction only.

Calculation of the pressure in a liquid.

To calculate the pressure at a point in a liquid, find the weight of the column of liquid standing on unit horizontal area at that depth. This will give the pressure not only downwards, but sideways, upwards and in any direction.

We can show by this method that at a depth h cm., in a liquid of density d gm. per c.c., the pressure, p gm. per sq. cm., is given by the simple equation

$$p = hd.$$

Imagine a horizontal area of 1 sq. cm. at a depth h cm. in the liquid (see Fig. 17).

Volume of liquid standing on 1 sq. cm. $= h$ c.c.

Weight of liquid standing on 1 sq. cm. $= hd$ gm.

∴ Pressure of liquid $= hd$ gm. per sq. cm.

Other units, such as lb. and ft., may of course be used.

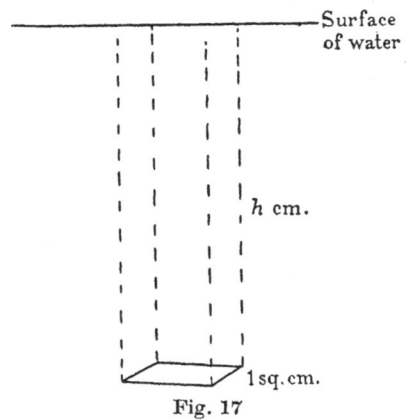

Fig. 17

To find the pressure on a diver at a depth of 100 *ft. in sea water.* Suppose the diver's helmet to be exactly 100 ft. below the surface of the water. Each horizontal sq. ft. of his helmet will support a column of water 100 ft. high and 1 sq. ft. cross-section.

Volume of water pressing
down on 1 sq. ft. of diver $= 100$ cu. ft.

Density of sea water $= 64$ lb. per cu. ft.

∴ Weight of water pressing
down on 1 sq. ft. of diver $= 100 \times 64$
$= 6400$ lb.

Thus pressure on the diver
due to the water $= 6400$ lb. per sq. ft.

Additional pressure on the diver due to the atmosphere. We shall see in the next chapter that the atmosphere exerts a pressure of about 15 lb. per sq. in. Thus the value we obtained for the pressure on the diver must be increased by 15 lb. per sq. in.

Pressure on the diver due to water $= 6400$ lb. per sq. ft.

$$= \tfrac{6400}{144} \text{ lb. per sq. in.}$$

$$= 44\tfrac{4}{9} \text{ lb per sq. in.}$$

$\therefore$ Total pressure on the diver due to

the water and the atmosphere $= 44\tfrac{4}{9} + 15$ lb. per sq. in.

$$= 59\tfrac{4}{9} \text{ lb. per sq. in.}$$

Deep-sea diving.

In recent years treasure worth many millions of pounds has been rescued from the bottom of the sea. Since an ordinary person cannot hold his breath for longer than half a minute without discomfort, or for longer than three minutes without dying, the divers engaged upon this work had to wear diving suits in which the air was continually replenished.

Now the pressure in the sea increases steadily with the depth. At the great depths in the Pacific, the pressure is so tremendous that thermometers sent down by the research ship *Challenger* were ground to powder.

The greatest depth at which a diver in a rubber suit can work is about 120 ft. The air in the suit has to be compressed to the same pressure as the water outside, as otherwise the pressure of the sea would squeeze the air out of it. British Admiralty divers wearing rubber diving suits salvaged £5,000,000 worth of bullion from the liner *Laurentic*, which lay in 120 ft. of water off the coast of Northern Ireland, having been sunk by a mine in 1917.

An unpleasant disease known as "bends" or "compression disease" attacks divers who work at such high pressures. On coming to the surface they experience acute pains in their joints and sometimes they become blind or unconscious, owing to the nitrogen in the air, which has dissolved in the blood at high pressure, bubbling out when the pressure is removed. Accordingly divers are hoisted slowly to the surface with rests to allow of adjustment to lower pressure at several depths, or placed in a chamber containing compressed air, where the pressure is gradually reduced over a space of several hours.

By courtesy of "The Times" and Messrs Faber and Faber

Fig. 18. A diver in a steel diving suit being lowered into the sea.

During the last ten years a steel diving suit has been perfected which is so strong that it can be used at a depth of 400 ft., and the air pressure inside need not be greater than normal. The arms and legs move on ball-bearing joints covered by watertight rubber membranes rather like eyelids. The suit is so heavy that the diver has very little freedom of movement; he can scarcely

By courtesy of Wide World Photos

Fig. 19. Dr Beebe's steel sphere in which he descended over 2000 feet below the surface of the sea. His assistants are here seen fixing up his telephone immediately prior to his descent.

walk, and can only grip things clumsily with pincers, rather like scissors, which are operated by handles inside the arms.

The most wonderful achievement of deep-sea diving was the salvaging of the *Egypt's* gold by the Italian ship *Artiglio* in 1932 and 1933, after four years of incredible patience and persistence.

In 1922, the liner *Egypt* sank after collision with a French steamer in a fog, about 40 miles from Brest, in the Bay of Biscay.

She had on board 5 tons of gold, and 43 tons of silver in bars and coin, worth over £1,000,000.

She settled on the bottom in a depth of 400 ft. of water, under a pressure of water of 190 lb. per sq. in.

Experts regarded salvage as impossible, owing to the great depth and terrible pressure, and also to the continual storms and underwater currents which are prevalent in that region.

By courtesy of Wide World Photos

Fig. 20. Dr Beebe emerging from his steel sphere after a descent.

The Italian divers descended in steel observation shells, built to withstand the great pressure from the outside, while inside they breathed air at ordinary atmospheric pressure. The shell had no arms or legs, only five windows made of glass an inch thick, like goggle eyes, through which the diver peered as he directed by telephone the movements of the tearing grabs and bomb charges let down from the deck of the *Artiglio*.

The wreck had first to be located by dragging the ocean floor with a sweep wire (during which process a number of other wrecks were found), and its position marked with buoys. Four decks were cut through or blown up to reach the bullion room, and the treasure drawn up by steel grabs. All this was done by machinery operated by men on the deck of the *Artiglio*, who were directed by divers swaying about in the gloom 400 ft. below them, under a mass of water as high as the spire of Salisbury cathedral.

But man has descended far deeper into the sea than 400 ft. In 1932 Dr Beebe, the American naturalist, was let down in a steel ball 4 ft. 9 in. in diameter, and 1½ in. thick, to a depth of 2200 ft. Through thick glass windows he could discover in the inky blackness strange phosphorescent fishes, hardly anything more than "a nightmare mouth" which allowed the water to enter inside them, and so counteract the external pressure.

Diving bells.

The foundations of breakwaters and the piers of bridges have often to be laid under water. In the case of breakwaters workmen are lowered on to the sea bed in a diving bell. This consists of a heavy steel box without a bottom, into which compressed air is blown to prevent water from entering. That used in constructing the breakwater at Dover (when the harbour was extended) weighed 35 tons, and its dimensions were 17 ft. × 10 ft. × 6½ ft. While being lowered and raised, the workmen stood on a small platform inside the bell (see Fig. 21 (b)). As they prepared the ground for the 50 ton concrete blocks of which the breakwater was built the air inside the bell had to be kept pure. They were in constant telephonic communication with the surface, and the bell was illuminated with electric light.

Let us suppose that the depth at which they were working was 30 ft. We can calculate the necessary pressure of the air in the diving bell.

Depth of water = 30 ft.
Density of sea water = 64 lb. per cu. ft.
∴ Pressure due to 30 ft. of water = 64 × 30 lb. per sq. ft.
$$= \frac{64 \times 30}{144} \text{ lb. per sq. in.}$$
= 13·3 lb. per sq. in.

Since the pressure of the atmosphere on the surface of the water is 15 lb. per sq. in., the total pressure in the diving bell

By courtesy of *Thomas Nelson and Sons*

Fig. 21 (*a*). Diving bells used in the Dover Harbour works. Two can be seen clearly, one in the right foreground and the other in the middle background. They are suspended from Goliath cranes.

will have to be $15 + 13 \cdot 3 = 28 \cdot 3$ lb. per sq. in., i.e. nearly double ordinary atmospheric pressure.

In case the air-compressor pump should break down, the air pipe has a valve opening only downwards. The bell, therefore, could not suddenly fill with water, and could be raised in time.

When laying the foundation of a bridge pier, a special type of diving bell known as a *caisson* is used. It consists of a steel

By courtesy of Thomas Nelson and Sons

Fig. 21 (*b*). The interior of a diving bell used in the Dover Harbour works.

cylinder weighted with concrete, and ultimately becomes part of the bridge.

Fig. 22 is a diagrammatic representation of a caisson. The central shaft and the lower part of the caisson are filled with compressed air. The men dig away the river bed until the cylinder has sunk to a solid foundation, when the whole is filled with concrete. The ladder by which the men descend the air pipes, and the apparatus by which the earth is lifted out, are not shown in the diagram. However, a simple representation of an airlock by means of which workmen enter the compression chamber is shown.

The workman enters a small room at the side of the central shaft, closing an airtight door behind him, and compressed air is blown in at the same pressure as that in the caisson. He then opens another airtight door, and enters the shaft. In this way he can enter the caisson without endangering others working below.

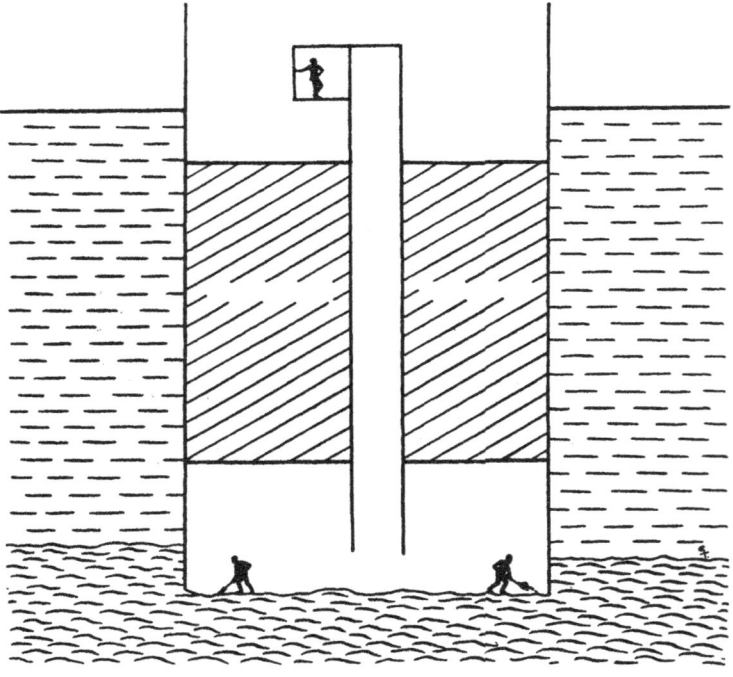

Fig. 22. Caisson.

Caissons were used in the construction of the foundations of two of the towers of the Forth Bridge in Scotland, one of the great bridges of the world. They were over 90 ft. high and had a diameter of 70 ft. at the bottom tapering to 60 ft. at the top. As each was towed to the spot where it was to be sunk, it looked like a small gasometer. When it had been weighted with concrete and lowered, the mud and ooze was sucked up from the river bed by powerful pumps. Then workmen descended and dug for three months, causing a gradual subsidence of the caisson. Finally it was filled with concrete. Each of the towers of the bridge rests on four such piers. (See p. 207 for picture of bridge.)

The hydraulic press.

The hydraulic press is an instrument for producing a large force from a small one. It consists, in its simplest form, of a narrow and a wide cylinder fitted with pistons, containing water, and connected by a tube (see Fig. 23).

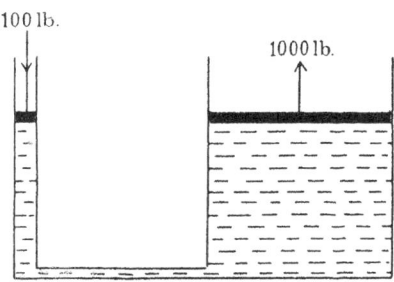

Fig. 23

Let us suppose that the areas of cross-section of the two cylinders are 1 sq. in. and 10 sq. in., respectively, and that a man pushes down on the small piston with a force of 100 lb. The pressure applied to the water is 100 lb. per sq. in., since the area of the small piston is 1 sq. in. The water transmits the pressure to the under side of the large piston (of area 10 sq. in.), and thus exerts a total force on it of $100 \times 10 = 1000$ lb.

The 100 lb. force has been magnified 10 times, and the machine is said to have a mechanical advantage of 10. Generalising, we can say,

$$\text{Mechanical advantage} = \frac{\text{Area of large piston}}{\text{Area of small piston}}.$$

Nature, however, never allows us to get something for nothing. The increase in the force must be, as it were, paid for. When the smaller piston is pushed down 1 in., the larger piston rises only $\frac{1}{10}$th of an inch, for when 1 cu. in. of water is pushed out of the small cylinder into the large one it has to spread out over an area of 10 sq. in. On the side of the larger force there is a smaller rise.

The hydraulic press was invented by Bramah, a Yorkshireman, in 1795. At first he had great difficulty in preventing the water from squirting out between the pistons and the cylinders, without making them fit so tightly as to ruin the efficiency of the machine. His foreman, Maudslay, suggested a very neat way of overcoming the trouble. A slot was cut out of the walls of each cylinder, and a circular leather washer with a cross-section shaped like an inverted letter U fitted into it so that the pressure of the water forced it tightly against the cylinder and the piston (see Fig. 24). The greater the pressure of the water, the more tightly

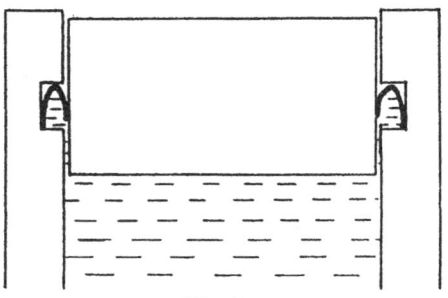

Fig. 24

was the washer pressed against the cylinder and the piston, and the more watertight the joint.

Jacks for lifting motor cars often take the form of a small hydraulic press. The liquid used is oil instead of water. With the very simple press shown in Fig. 24, as soon as one stopped pushing on the small piston the car would descend with a bump. Consequently a valve opening only from the smaller to the larger cylinder is fitted in the tube to prevent the liquid running back (see Fig. 25). But if this valve alone were fitted, once the car had been raised it could never be lowered. Hence another tube with a tap connects the cylinders to enable the liquid to flow back. Fig. 25 is, of course, purely diagrammatic. The hydraulic jack is not this shape, and a long lever is usually fitted to enable quite a large force to be applied conveniently to the small piston.

Huge hydraulic presses capable of delivering blows of 6000

tons at the rate of 40 per minute, or even as great as 12,000 tons, are used for forging steel.

Certain metal parts of a machine on which there is not excessive strain can be made by pouring molten metal into a mould. Parts made in this way are known as castings. The mould is made of sand of the right size and shape by packing it round a wooden model. But those components of a machine which require to be very strong must be forged, that is, they must be hammered out from white-hot lumps of metal.

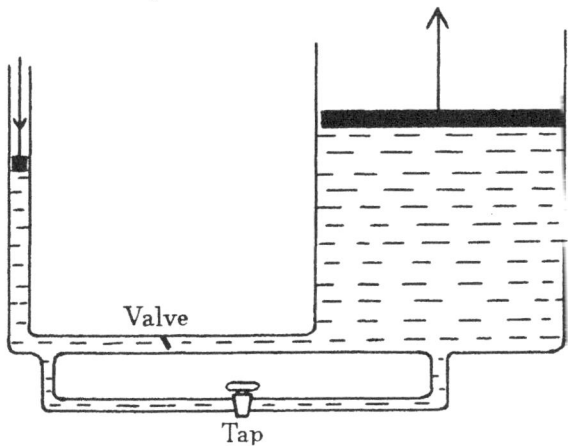

Fig. 25. Simplified diagram of car jack.

Fig. 26 is a picture of a large press used for forging steel. The press, relative to the simple one described above, is upside down. The upward force on the small piston is produced by coupling it up to a special steam engine, and the large piston is connected to a huge hammerhead, shaped according to the casting it must make, which, guided by steel pillars, descends vertically. White-hot iron ingots are held in an anvil beneath the hammer, and formed, let us suppose, into a liner's propeller shaft, or bent into armour plating for a battleship.

You should endeavour to be taken round a works with such a press, for its operation is awe-inspiring. The hammer descends with sickening thuds, causing showers of sparks, and the steel is shaped before one's eyes.

The hydrostatic paradox.

The most interesting and important experiments on liquid pressure were performed by a remarkable Frenchman, Blaise Pascal (1623–62). A man of outstanding intellect, he had made his name as a mathematician by the age of sixteen, but at thirty-three he renounced science for the religious life. He is best known as one of the greatest of French prose writers.

By courtesy of the English Steel Corporation, Ltd.

Fig. 26. A large hydraulic press. The white-hot steel is being forged: while it is slowly revolved the hammerhead descends repeatedly with great force.

Pascal demonstrated that the pressure in a liquid is proportional to the depth. He also showed that the pressure in a liquid can act in all directions, and that *it exerts a force at right angles to any surface placed in it.*

In Fig. 27 columns of liquid of equal height, but in vessels of different shape, are shown. It is not at once apparent, in this case, that the pressure or force per unit area at the foot of each column must have the same value, since the walls of the vessels are not vertical.

Nevertheless, the pressures at the bottom of the columns are equal, since the water would not flow from one to the other if they were connected. We deduce that *the pressure of the liquid in a vessel is not affected by the shape or cross-section of the vessel.*

Suppose the three vessels in Fig. 27 have the same base area. We know that the pressure on the base of each is the same. Hence the force on the base of each is the same, since they all have equal areas.

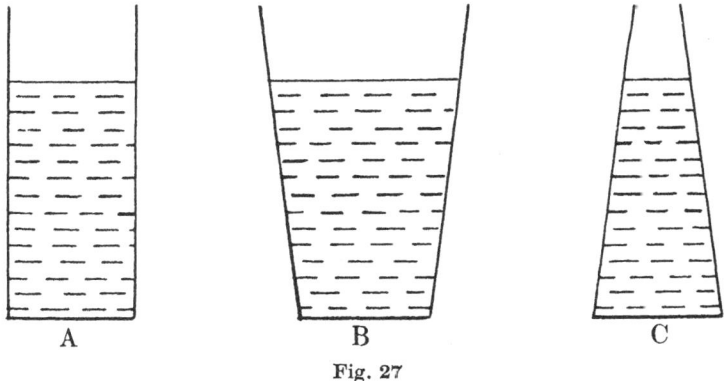

Fig. 27

Now the force on the base of A is obviously exactly equal to the weight of water in A. The force on the base of B must therefore be less than the weight of water in B, and the force on the base of C *greater* than the weight of water in C. This remarkable conclusion, known as the hydrostatic paradox, was first explained by Pascal.

In the case of B, the sloping walls support some of the weight of the water. But how are we to explain the case of C? The water presses at right angles to the sloping walls of C and hence tends to push them upwards. The walls exert an equal downward pressure (action and reaction are equal and opposite, see p. 117) and this is transmitted through the water to provide the extra force on the base.

The equal forces on the bases of *A*, *B* and *C* can be demonstrated experimentally by making the bases detachable and supporting them on balances. The walls of the vessels, of course, must have separate rigid supports.

An experiment to compare the densities of two liquids.

If a U-tube contains water, the surface levels in each arm are equal. If, however, the vessel contains two different liquids which do not mix, the heights of the balancing columns will be found to be different, depending on the densities of the liquids.

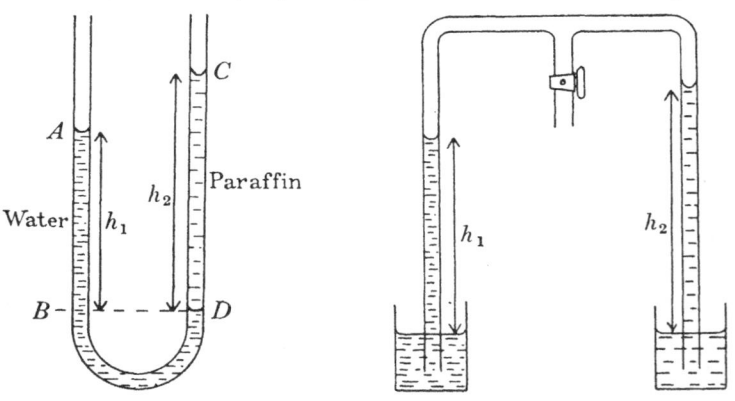

Fig. 28   Fig. 29. Hare's apparatus.

Pour two liquids which do not mix, such as water and paraffin, into the arms of a U-tube (see Fig. 28).

The column of water *AB* is balancing the column of paraffin *CD*. Thus the pressures due to these two columns of liquid must be equal. Measure the heights of the columns h_1 and h_2. Suppose the densities of the liquids are d_1 and d_2. Then,

Pressure due to column *AB* = Pressure due to column *CD*,

i.e. $$h_1 d_1 = h_2 d_2;$$

$$\therefore \quad \frac{d_1}{d_2} = \frac{h_2}{h_1}.$$

If the two liquids mix, an inverted U-tube is used, commonly in the form of Hare's apparatus (Fig. 29). By opening the clip at A and sucking out air, the liquids may be caused to rise to any desired height, and remain there when the clip is closed.

Measure h_1 and h_2.

The pressure of the atmosphere supports the liquid columns, and since the pressure of the air above both columns is the same, the pressures exerted by the columns must be equal.

As before $$\frac{d_1}{d_2} = \frac{h_2}{h_1}.$$

The manometer. Determination of the pressure of the gas supply.

We have seen that the pressure exerted by a liquid depends solely on its depth and density. It is therefore possible to measure a pressure such as that of the gas supply by balancing it against the pressure of a column of liquid of known density, and measuring the height of the column. We shall see in the next chapter that it is a common practice *to denote a pressure simply by the height of the column of liquid it can support,* without bothering to calculate its value in lb. per sq. in., gm. per sq. cm. or similar units.

Fig. 30 (*a*) represents a pressure gauge working on this principle, known as a manometer. It consists of a glass U-tube containing water. When open to the atmosphere, the level of the water in each arm is the same. When, however, one arm is connected to a gas tap by means of a piece of rubber tubing, and the gas turned on, a difference in level results. The pressure of the gas supply has forced the water down on the one side, and up on the other, until the column of water which constitutes the difference in water levels on each side is heavy enough to balance the pressure of the gas.

The difference in level of the water in the arms of the U-tube is measured, and gives, not the actual gas pressure, but how much greater the gas pressure is than atmospheric pressure. The atmosphere is pressing down on the surface of the water in the open arm, and its value must be determined by means of a barometer in order to obtain the total value of the gas pressure.

Fig. 30 (*b*) represents a water manometer with one arm much wider than the other. Would the difference in level in this case differ from that in (*a*)? Here we have a precisely similar problem to that at the beginning of this chapter, when we had to explain why water in wide and narrow tubes connected together remains at the same level in both. The pressure of a liquid in a vessel does

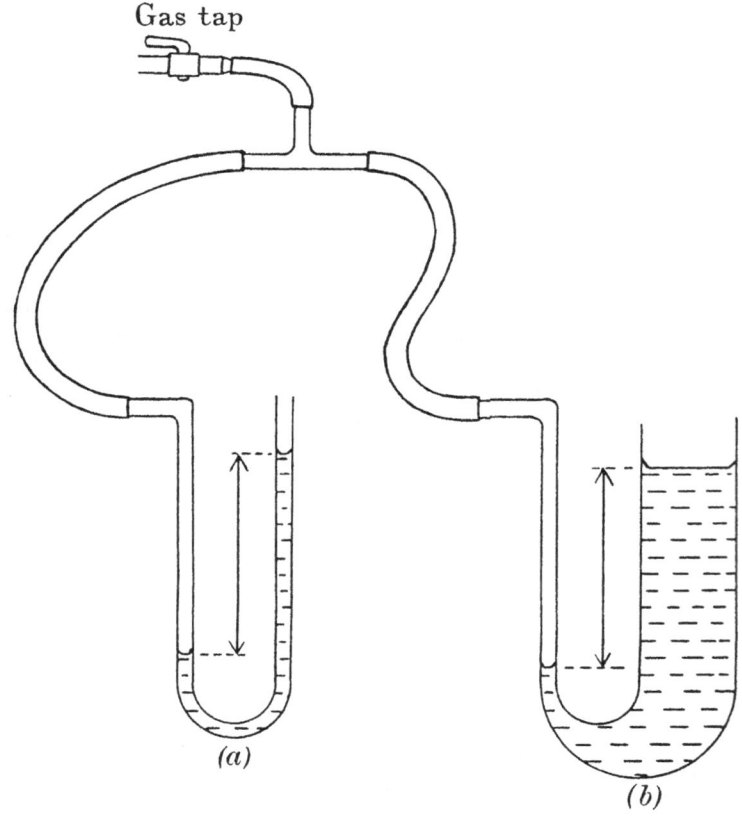

Fig. 30

not depend on its cross-section, since the greater weight in a wider tube has a greater area to act over. Hence the difference in level in manometer (b) will be exactly equal to that in manometer (a).

Calculation of total gas pressure.

(*If this calculation and that in the next section are found difficult to follow, they should be postponed until the account of the pressure of the atmosphere and barometer in Chapter* IV *has been read.*)

$$
\begin{aligned}
\text{Difference in level in water manometer} &= 6 \cdot 0 \text{ cm. of water.}\\
\text{Pressure of the atmosphere} &= 76 \cdot 00 \text{ cm. of mercury.}\\
\text{Density of mercury} &= 13 \cdot 6 \text{ gm. per c.c.}\\
\therefore 6 \text{ cm. of water} &= \frac{6}{13 \cdot 6} \text{ cm. of mercury}\\
&\equiv 0 \cdot 44 \text{ cm. of mercury.}\\
\therefore \text{Total gas pressure} &= 76 \cdot 00 + 0 \cdot 44\\
&= 76 \cdot 44 \text{ cm. of mercury.}
\end{aligned}
$$

Blowing over a brick.

An amusing experiment to illustrate *the difference between pressure (i.e. force per unit area) and force or thrust (pressure × area)* may be performed with a brick and paper bag.

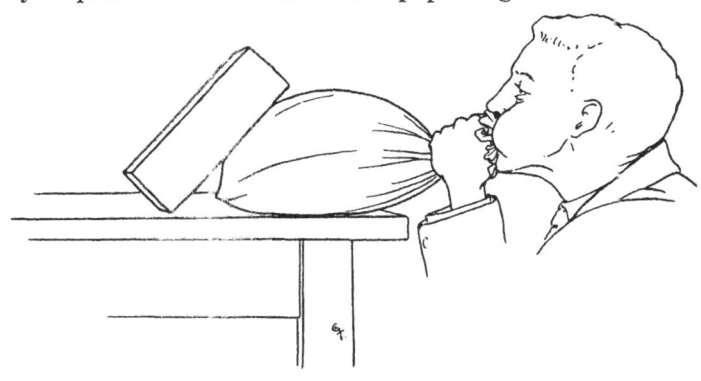

Fig. 31

A brick weighs about 10 lb., and no amount of blowing will cause it to overturn. By placing a paper bag against the side, and blowing into the bag, the brick can be overturned easily. It can also be raised up by blowing into a bag which has been placed underneath.

Let us perform an experiment to determine with what pressure a normal person can blow. A mercury manometer, i.e. a U-tube containing mercury, is required. Let one experimenter blow with all his power into one arm, while another experimenter determines the difference in level of the mercury in the two arms. The following are the results of such an experiment:

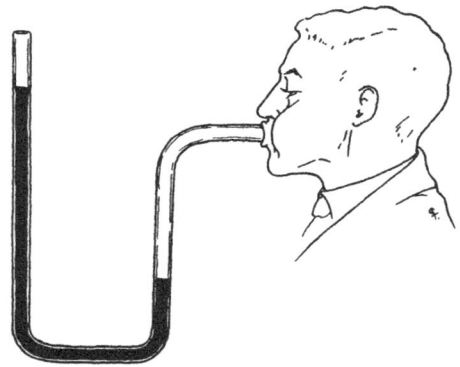

Fig. 32

Pressure with which a normal person can blow.

Difference in level of mercury = 6 cm.

We wish to convert this reading into lb. per sq. in.

Now pressure of atmosphere = 76 cm. of mercury,
also   = 15 lb. per sq. in. approx.

∴ 76 cm. of mercury   ≡ 15 lb. per sq. in.

1 cm. of mercury   ≡ $\frac{15}{76}$ lb. per sq. in.

6 cm. of mercury   ≡ $\frac{15}{76}$ × 6 lb. per sq. in.

                        ≡ 1·18 lb. per sq. in.

∴ Blowing pressure exerted by
experimenter   = 1·18 lb. per sq. in.

Calculation of force or thrust exerted by paper bag on brick to overturn it.

Excess pressure in bag over
that of atmosphere   = 1·18 lb. per sq. in.

Area of flat base of paper bag = 20 sq. in.

Thrust   = Pressure × Area.

∴ Thrust exerted by bag on brick = 20 × 1·18

                              = 23·6 lb.

From this one can understand why the brick was blown over so easily.

The utilisation of water pressure.

Water wheels. A vast amount of untapped energy runs to waste unceasingly in the waterfalls and rivers of the world. Running water has been called "white coal". Less than one-quarter of the flow over the Niagara Falls is utilised for generating electricity, but nevertheless it is sufficient to supply 550 towns—work their factories, run their trains, and light their streets and houses.

Fig. 33. Undershot water wheel.

The old-fashioned way of utilising water pressure, or water power, is by means of a water wheel. If the stream runs swiftly a wheel with paddles is mounted so that the lower paddles dip into the stream and the moving water turns the wheel. This is known as an *undershot wheel* (see Fig. 33).

Sometimes water is fed into pockets (called buckets) on one side of a wheel, making that side heavier, and hence turning the wheel. In an *overshot wheel* (see Fig. 34) the water is fed on to the wheel near its highest point, and the water is retained in the buckets for one-third of a revolution. In a *breastshot wheel* (see Fig. 35), the water is supplied half-way down the wheel, and the buckets are shaped so that the water remains in them for about one-fourth of a revolution.

The overshot and breastshot wheels, worked by the weight of

falling water, are more than twice as efficient as the undershot wheel, but they are convenient only when a stream drops suddenly, say over waterfalls, and water can be led away in a special channel from the top of the falls to a wheel.

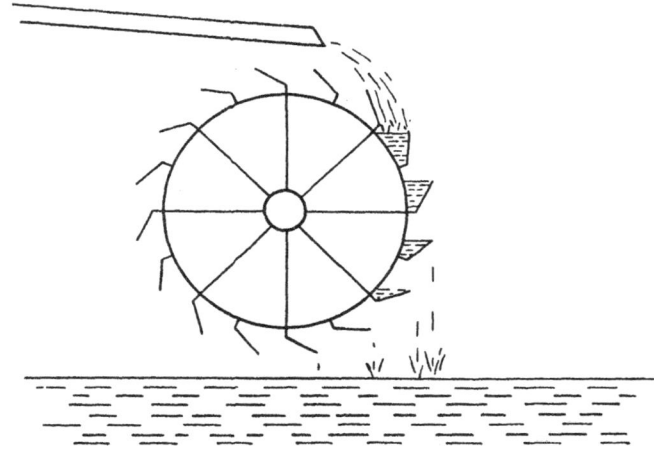

Fig. 34. Overshot water wheel

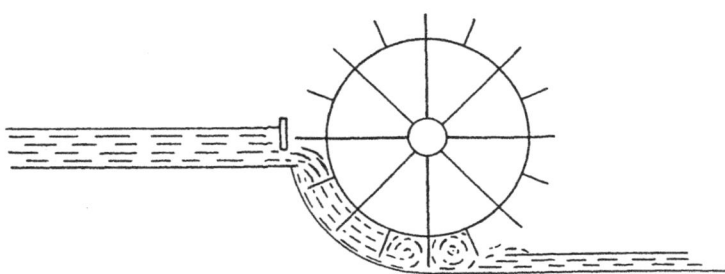

Fig. 35. Breastshot water wheel.

The rivers of Great Britain are suitable for water wheels, and can generate up to 50 horse-power. However, as coal is plentiful, water power has been replaced largely by steam engines, and derelict water mills are to be found all over the country.

By courtesy of the Swiss Federal Railways

Fig. 36. The dam and reservoir of the Barberine power station, Switzerland (see Fig. 37).

By courtesy of the Swiss Federal Railways

Fig. 37. The Barberine power station. The water drops through the pipes on the hillside from the reservoir shown in Fig. 36.

Turbines. In countries like Switzerland and Norway, the rivers fall thousands of feet, and are capable of producing an enormous horse-power. Here, highly specialised water wheels known as turbines are used.

Usually a dam has to be built, creating a large reservoir high up in the mountains, from which water is led away through a

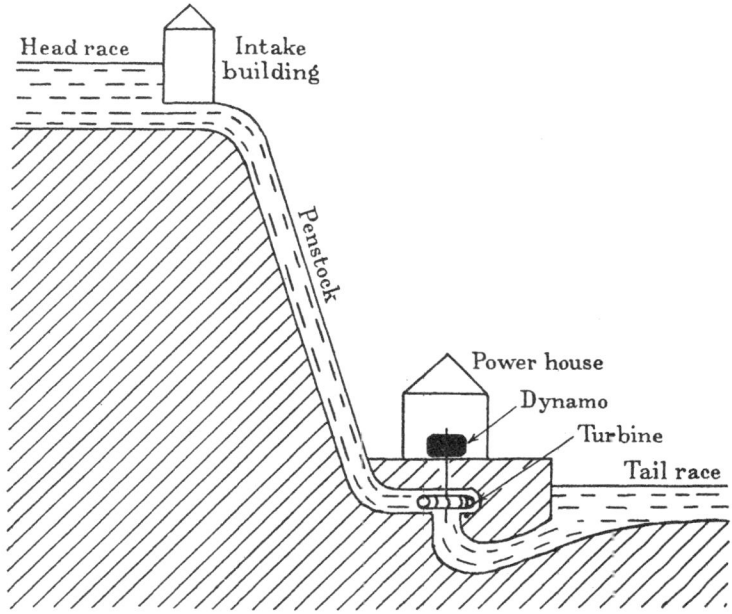

Fig. 38. Diagram showing how water reaches the turbine, which turns the dynamo in the power house.

channel on high ground called the *head race,* to where it can drop through steel pipes called *penstocks.* At Fully, in Switzerland, the water drops 5320 ft.—one of the largest heads in the world.

The water then passes through a turbine, a machine which must be of enormous strength, and this is coupled to a generator for producing electricity. Having given up all its energy, the water flows away through a channel called the *tail race* (see Fig. 38).

The power available depends on two things: (i) the quantity of water, called the *flow*, (ii) the height through which it falls, called the *head*.

There are two types of turbine: (i) *for low heads and large flow, the Francis or reaction turbine,* (ii) *for high heads and small flow, the Pelton wheel or impulse turbine.*

The reaction turbine consists of a pipe (which may be wider than a railway tunnel) curled round like a snail so as to lead the

By courtesy of Messrs Boving and Co., Ltd.

Fig. 39. The spiral casing which leads the water into the runner of a reaction turbine. The runner (shown in Fig. 40) fits into position where the upper man is standing.

water round a circular shell or casing. In the middle is a massive wheel, mounted on a vertical shaft. The problem is to use the pressure of the water, passing through in great bulk, to turn the wheel.

Fig. 40. The runner of a reaction turbine. Water passes between the curved vanes, causing the runner to rotate, and then drops out through the hollow middle.

The wheel is therefore equipped with vanes (see Fig. 40) which do not point to the centre, but are inclined, and the outer shell has fixed guide blades inclined from the centre in the opposite way.

Water passing from the outer shell sideways into the spaces between the vanes gets slewed round and consequently gives the wheel a shove. It then drops vertically out of the wheel. Since the flow of water is continuous and is hitting all the vanes at once, it exerts a steady force on the wheel all the way round its

By courtesy of Messrs Boving and Co., Ltd.

Fig. 41. A Pelton wheel. A jet of water, under very high pressure, is played on to the "buckets" causing the wheel to revolve. The man is holding one of the buckets vertically to show its shape.

circumference. The amount of water flowing through the great turbines on the river Shannon is 100 tons per sec.

The Pelton wheel is a massive wheel with buckets of a special shape, on to which one or more jets of water are played causing it rapidly to revolve. This is the type of turbine used at Fully, where the head of water, as mentioned above, is 5320 ft. The jet of water as it strikes the buckets is $1\frac{3}{8}$ in. in diameter, and has a velocity of 590 ft. per sec. Such a jet would knock a wall down or reduce the human body to pulp. A Pelton wheel is shown in Fig. 41. The buckets are made curved as shown, so that the jet strikes them without splash or shock; and the speed of the wheel

By courtesy of Canadian National Railways

Fig. 42. An aerial view of the Niagara Falls. The drop over the Falls is about 160 feet. Several power stations utilise part of the flow for generating electricity.

is kept at half that of the jet, as this enables the water to give up all its energy.

Power from Niagara. The Niagara river joins Lake Erie to Lake Ontario and falls 326 ft. in a distance of 33 miles. About 160 ft. of this drop takes place in the Niagara Falls themselves. It has been estimated that about 8000 tons of water on an average drop over the Falls in each second. A treaty has been signed between Canada and the United States to allow only one-quarter of this flow to be used for power, in order to preserve the spectacle. In 1895 a 15,000 horse-power station was built on the United States side of the Falls. On the Canadian side the Toronto and Ontario companies between them develop 350,000 horse-power.

These power plants utilise a head of about 160 ft.—the drop over the Falls—but the latest plant, that of the Queenston-Chippawa Power Development, utilises 310 out of the 326 ft. drop between the Great Lakes. Two miles above the Falls water is led away along the channel of the old Welland river, whose flow was reversed, and through a canal over high ground to a site where the water drops through steel pipes 383 ft. long and 14 ft. to 16 ft. wide. There are nine reaction turbines working at $187\frac{1}{2}$ revolutions per minute, and each unit (turbine and generator) weighs 1044 tons with rotating parts weighing 340 tons. The total horse-power developed is over half a million.

Harnessing the Shannon. The largest hydro-electric plant in the British Isles is in Ireland, where the 100 ft. drop of the river Shannon between Killaloe and Limerick is used to generate 34,000 horse-power. The scheme was devised by the Irish engineer, Dr McLaughlin, and the contractors were the German firm of Siemens. The work was completed a few years ago, and has proved a great success.

Dams.

Among the most remarkable engineering feats must be ranked the building of great dams. Indeed, the dams on the Nile are as splendid a monument to British engineers as are the Pyramids to the ancient Egyptians.

Dams are required (i) for power schemes—great artificial lakes have been made in the Alps to provide electricity for the Swiss railways; (ii) to create reservoirs for the drinking supply of a city—the Derwent dam in the North of Derbyshire has created a lake to supply several towns with water; (iii) for irrigation—this is the purpose of the great dams in Egypt and India.

Dams are always made thicker at the base than at the top (see Fig. 43), since the pressure of water increases with depth. They are built of masonry and require a deep foundation; otherwise they may be swept away bodily. They must be absolutely

watertight, since a small leak would weaken the whole structure.

Sluice gates are often built in the dam to allow water to pass
through when the river or reservoir gets too full. At Lake
Vyrnwy, which supplies Liverpool with water, the excess water
overflows like a waterfall over the crest of the dam.

Terrible consequences may ensue if a large dam breaks. Some
years ago a dam gave way in Pennsylvania. A great wall of
water rushed down upon the town of Johnstown, reducing it to
ruins and drowning 10,000 people.

Dams on the Nile. Egypt has practically no rainfall, and depends for its water on the river Nile, which draws its supply from
the tropical rains of the Nyanza basin, and the melting snows of
Abyssinia, 2000 miles upstream. The whole country has to be

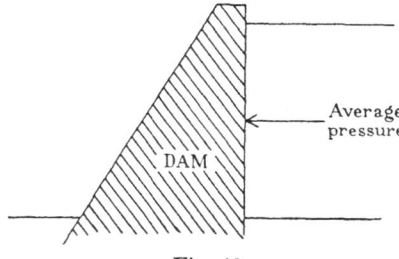

Fig. 43

irrigated by an intricate system of canals. Sir Benjamin Baker,
the engineer responsible for raising the Assouan dam, said: "It
will be seen what a vast amount of labour is saved throughout
the world by the providential circumstance that in ordinary cases
water tumbles down from the clouds and has not, as in Egypt,
to be dragged up from channels and wells".

The real problem is that the Nile does not give a steady supply
of water. Its waters begin to rise in June, and continue to rise
until mid-October, when it is racing in full flood. It then recedes
slowly until its waters are very low indeed. Thus during half the
year there is too much water, and during the other half, not
enough.

For long ages, men have realised what an advantage it would
be if the Nile's waters could be prevented from running to waste

into the Mediterranean during flood time and stored up for use during its ebb, but it is only in modern times that they have had the ability to cope with the task.

By courtesy of Sir M. Macdonald and Partners

Fig. 44. The Assouan dam on the Nile. During flood time the river flows through the sluice gates but during its ebb the latter are closed and the water is dammed up for irrigation purposes.

About 100 years ago French engineers built a barrage across the Nile below Cairo, which was strengthened by British engineers in 1890. Other low barrages at Zifteh, Isna and Assiout enable most of Lower Egypt to be irrigated.

But the largest dams are higher upstream at Assouan and

Sennar, and each of these is capable of holding up several thousand million tons of water.

At Assouan is the first of the cataracts of the Nile and when the river was low it flowed in channels between rocky islands. The banks are high cliffs, and this is an ideal site for building a dam to raise the level of the river.

The work commenced in 1898 and a year was spent in assembling the necessary machinery and building a temporary town for the workmen. The method of construction was to dam up each channel between the islands in turn, by dropping great blocks weighing several tons into the stream above and below the opening. The enclosure so formed was called a sudd, and when it was complete the water was pumped out and the workmen commenced digging the foundations. Work was impossible during the seven months of flood time, and so during the rest of the year the construction proceeded night and day, at night under the illumination of large arc lights.

The dam is $1\frac{1}{4}$ miles long, 120 ft. high above the deepest foundations, 16 ft. thick at the top and 100 ft. thick at the bottom. It has 180 sluice gates. It was not built as high as the original designs owing to the great public outcry when it was learned that the plan entailed the submerging of the island of Philae, on which stands an ancient Egyptian temple. However, the dam proved such a source of wealth that this objection was over-ridden. In 1907–12 it was raised 16 ft. and thickened, and at the present time (1934) it is being raised still another 27 ft.

The Sennar dam across the Blue Nile, completed in 1925, is the largest in the world. It is about two miles long, and holds back water for 58 miles. It has converted a vast desert into a fertile country for the growing of cotton.

A similar tale could be told of India. The Tata power works, the Sukkur barrage and the Mandi hydro-electric scheme are great achievements—monuments to the grandeur of the civilisation of the West.

The force of water on the Assouan dam. The following figures are simplified, but are based on the dimensions of the Assouan dam:

> Height of dam = 100 ft.
>
> Length   ,,   = 2000 yd. = 6000 ft.
>
> ∴ Area   ,,   = 6000 × 100 = 600,000 sq. ft.

The pressure on the dam steadily increases with the depth. Thus we must find the average pressure, which is the pressure half-way down, i.e. at a depth of 50 ft.:

Density of fresh water $= 62\frac{1}{2}$ lb. per cu. ft.

$\therefore$ Pressure at a depth of 50 ft. $= 62\frac{1}{2} \times 50$

$= 3125$ lb. per sq. ft.

This represents the weight of a column of water of height 50 ft., standing on 1 sq. ft. The pressure of course acts sideways on the vertical face of the dam.

$\therefore$ Total force on dam $= 3125 \times 600,000$ lb.

$$= \frac{3125 \times 600,000}{2240} \text{ tons}$$

$= 837,000$ tons.

No account was taken in this calculation of the water on the other side of the dam. When the Nile is in flood, the levels on the two sides will be nearly the same, and consequently the resultant force on the dam due to the depth of the water will be nearly zero. There is, however, the force due to the velocity of the water, which is about 15 m.p.h. The sluice gates are then full open.

SUMMARY

$$\textbf{Pressure} = \frac{\textbf{Force}}{\textbf{Area}}.$$

Force or Thrust = Pressure × Area.

The pressure in a liquid depends on its depth and density but not on its cross-section, and can act in all directions. In order to calculate the pressure at a point in a liquid the weight of the column of liquid standing on unit horizontal area at this depth must be found:

$$p = hd.$$

Pressure is often measured in terms of the height of a column of liquid: a simple type of pressure gauge consists of a U-tube containing a liquid, and is called a manometer.

The densities of two liquids may be compared by balancing them in the arms of a U-tube, when

$$\frac{d_1}{d_2} = \frac{h_2}{h_1}.$$

A hydraulic press is a machine used for producing a large from a small force, in which water pressure is transmitted from a small piston to a large one.

$$\text{Its mechanical advantage} = \frac{\text{Area of large piston}}{\text{Area of small piston}}.$$

The pressure of liquids can be converted into water power by (i) water wheels, (ii) turbines.

Water wheels may be of three types: (i) undershot, (ii) breast-shot, (iii) overshot wheels.

Water turbines are of two main types: (i) the reaction or Francis turbine, (ii) the impulse turbine, or Pelton wheel. Turbines usually convert their power into electricity.

Dams, which must be built wedge-shaped, are constructed to provide reservoirs for (i) drinking water supplies, (ii) hydro-electric power, (iii) irrigation.

QUESTIONS

1. Define pressure.

The sharp end of a nail has an area of $\frac{1}{2000}$ sq. in., and is held in contact with a block of wood. What pressure will it exert on the wood when struck by a hammer with a force of 4 lb. wt.?

2. Explain carefully, noting in particular the physical principles involved:

  (a) Water runs more slowly out of a tap upstairs than downstairs.
  (b) One type of petrol gauge in a motor car consists of a glass tube in which the level of the petrol can be seen. (Draw a diagram.)

3. A beaker contains mercury (density 13·6 gm. per c.c.). If the area of the bottom of the beaker is 30 sq. cm., and the depth of the mercury is 5 cm., find (a) the force (or thrust), (b) the pressure, of the mercury on the bottom of the beaker.

4. The water pressure at a tap is 75 lb. per sq. in. How high above the tap must be the water tower from which the water is supplied?

5. An Austin 12 motor car weighs 18 cwt. and the pressure in each of the four tyres is 30 lb. per sq. in. What area of each tyre must be in contact with the ground?

6. Paraffin is poured into a glass U-tube containing some water. The liquids do not mix, and the heights of the columns of paraffin and water above their common surface are 10 cm. and 8 cm. respectively.

Find the density of paraffin in gm. per c.c., explaining your method fully.

7. A boy can lift himself by standing on the hydrostatic bellows shown in Fig. 45 and pouring water into the narrow tube *A*. Explain fully why this is possible.

A boy of weight 10 stones stands on *B*, which has an area of 1 sq. ft. To what height must the water in the tube *A* stand above that in *B* to balance his weight? (Density of water = $62\frac{1}{2}$ lb. per cu. ft.)

8. (*a*) Find the total force of water on a dam 300 ft. long when the depth is 40 ft.

(*b*) Will the dam have to be made stronger if the area of the lake is increased?

(*c*) Why is a dam built thicker at the bottom than at the top?

9. A conical flask, and a glass tumbler with vertical sides, each have a base of area 50 sq. cm. and contain water to a depth of 12 cm. Calculate (*a*) the pressure, (*b*) the force of the water on the base of each. How does the latter compare with the weight of water in the vessels? Explain.

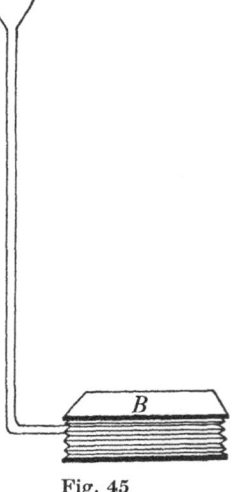

Fig. 45

10. Describe two forms of water wheel. Under what conditions should each be used?

11. Write an essay on the utilisation of water power.

School Certificate Questions

12. Describe experiments illustrating the properties of fluid pressure.

13. What is meant by the pressure at a point in a liquid? Describe an experiment to show how the pressure at a point in a liquid increases with the depth below the surface.

Calculate in dynes per sq. cm. the pressure at a point 100 metres below the surface of the sea on a day when the height of the mercury barometer at sea level is 745 mm. (Sp. gr. sea water, 1·03; mercury, 13·6. 981 dynes = 1 gm. wt.)

14. Describe the hydraulic press and explain its action. If a press is provided with cylinders of diameter 2 in. and 1 ft. respectively, what is its mechanical advantage?

15. A barrel which requires a pressure of 2 kg. to the sq. cm. to burst it is completely filled with water. A vertical tube 0·5 sq. cm. in cross-section projects vertically upwards from the top of the barrel. What weight of water must be poured into this tube to burst the barrel?

16. The free surface of water in a diving bell is 20 metres below the surface of the sea. The atmospheric pressure is 75 cm. of mercury. What will be the pressure of the air in the diving bell? (Sp. gr. of mercury and sea water are 13·6 and 1·025 respectively.)

Chapter IV

THE PRESSURE OF AIR

Air, although it is very light, has weight, and its density is about
0·0012 gm. per c.c. A room contains a much greater weight of
air than one might expect.

The following calculation gives the weight of air in an ordinary
classroom:

Dimensions of room	= 700 cm. × 600 cm. × 500 cm.
∴ Volume of room	= 210,000,000 c.c.
Density of air	= 0·0012 gm. per c.c.
∴ Weight of air in room	= 210,000,000 × 0·0012 gm.
	= 252,000 gm.
	= 252 kilograms
	= 500 lb. (approx.)
	= ¼ ton (approx.).

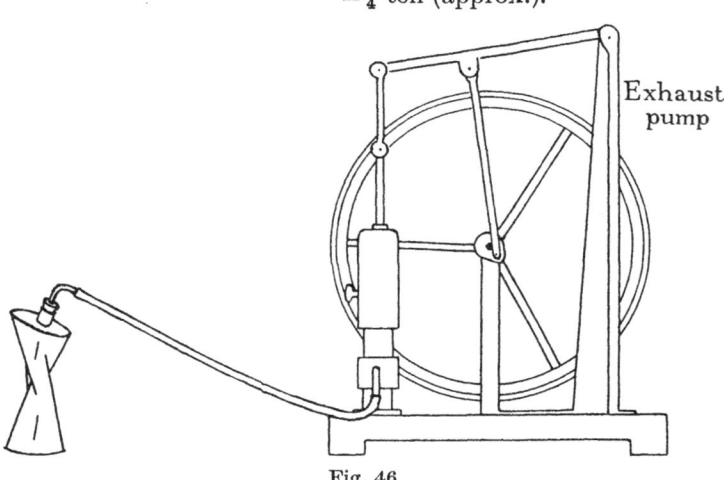

Fig. 46

When one considers what an enormous weight of air there
must be pressing on the earth, the following experiments will not
occasion much surprise.

Experiments to show the pressure of the atmosphere.

Experiment 1. A tin from which the air is extracted by means of an air pump crumples up. The pressure of the outside atmosphere is no longer balanced by the pressure of the air inside.

The pressure of the atmosphere is about 15 lb. per sq. in. It is interesting to calculate the number of square inches in the surface area of the tin, and so obtain the thrust or force of the air on the whole tin.

Surface area of tin used = 80 sq. in.
Pressure of atmosphere = 15 lb. per sq. in.
∴ Thrust or force on tin = 80 × 15 lb.
= 1200 lb.
= $\frac{1}{2}$ ton (approx.).

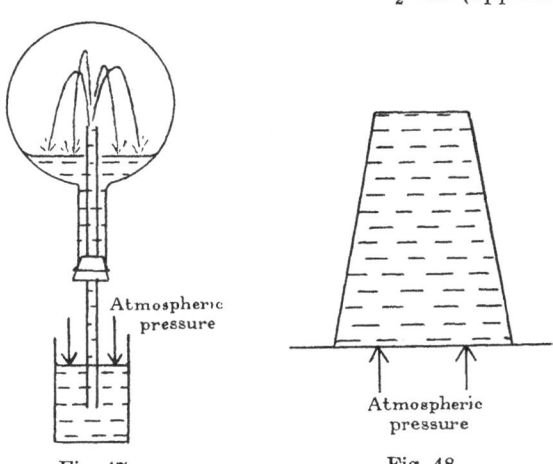

Fig. 47

Fig. 48

Experiment 2. A glass flask containing a little water is heated so that the water boils and the steam drives out the air. It is fitted with a cork through which passes a glass tube. On the end of the glass tube is a piece of rubber tubing, pinched by a clip. The flask is inverted, and the rubber tubing opened under the surface of water in a beaker. Water rushes up the tube, filling the flask, owing to the pressure of the atmosphere on the surface of the water in the beaker.

Experiment 3. A glass tumbler is filled to the brim with water, and a piece of cardboard placed on the top so that there are no air bubbles beneath it. The tumbler can now be inverted, and the water will not run out, since the pressure of the atmosphere under the card is greater than that of the water above it.

Experiment 4. A partially inflated balloon is placed in an air-tight bell jar which is connected to an air pump. As the air is drawn out of the bell jar, the balloon increases in volume, since the pressure of the gas inside is greater than that of the air in the bell jar. This experiment was first performed with a bladder by Robert Boyle, who thereby demonstrated that air exerts a pressure. He called it the "spring of air".

Historical. The vacuum.

For some years scientists were very interested in the problem of trying to obtain a vacuum, i.e. a perfectly empty space. It was during the course of these experiments that they first became conscious of atmospheric pressure.

Until the middle of the seventeenth century people believed that it was impossible to obtain a vacuum, since, in their experience, an empty space was always immediately filled. For instance, when the nozzle of a simple pump or syringe is placed under water, and the piston is raised, water rushes in to fill the partial vacuum left below the piston. But it was mainly the following remarkable experiment which led to the suspicion that Nature would never allow a vacuum to be made.

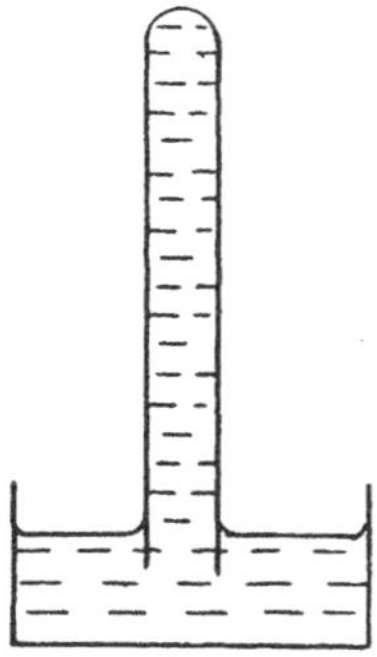

Take a long glass tube sealed at one end, and fill it absolutely full of water. Place the thumb over the open end, invert the tube, and remove the thumb under the surface of water in a bowl. The tube remains full of water, which does not run out as one might expect (see Fig. 49).

Fig. 49. The water will not run out of the tube.

People began to say "Nature abhors a vacuum". It was suggested that the water could not run down because it would leave a vacuum at the top, and this Nature would use all her power to frustrate.

Now there lived in Germany during the time of the Thirty Years' War a man of considerable ingenuity and resource, by name Otto von Guericke. He was very interested in this problem of creating a vacuum, and he devised a number of most interesting and spectacular experiments.

British Museum Photo: Fleming

Fig. 50. Von Guericke's experiment with the copper globe. His first globe, on being evacuated, "suddenly with a loud clap and to the terror of all" collapsed. This picture shows a stronger globe. Great force was needed to work the pump owing to the pressure of the atmosphere on the outside of the piston.

He first attempted to make a vacuum by filling a stout wine cask full of water to drive out the air, and then pumping out the water. He was amazed at the force required to work his pump. By the time he had drawn a large part of the water out of the cask, it required three strong men to pull out the piston of the pump, and before he could complete the experiment the seams of the cask gave way. There was a great hissing and bubbling as the air rushed in.

He then replaced the cask with a large copper globe, filled it with water to drive out all the air, and began to pump out the water. This time the whole globe collapsed with a loud report.

The supporters of the "horror vacui" theory could certainly point to pretty drastic action on the part of Nature in preventing the production of a vacuum.

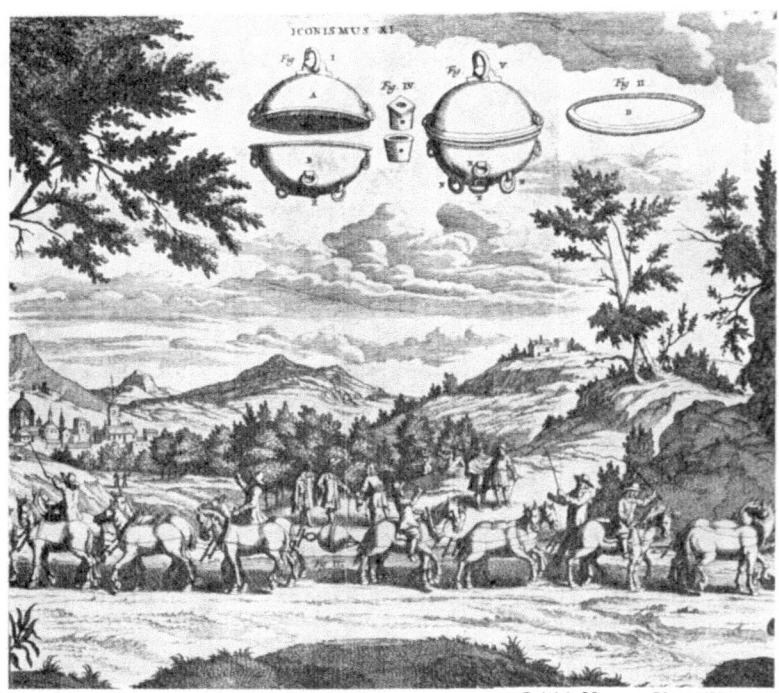

British Museum Photo: Fleming

Fig. 51. The Magdeburg hemispheres experiment. Two teams, each of eight horses, were needed to pull the evacuated hemispheres apart.

Von Guericke, however, realised that the real difficulty was the pressure of the atmosphere. Of many experiments which he devised to illustrate this, probably the most famous was one in which he used two copper hemispheres, now known as the Magdeburg hemispheres. (Von Guericke was burgomaster of

Magdeburg.) The hemispheres fitted together so perfectly that with a greased leather washer between them they made an air-tight joint. They could be pulled apart quite easily when full of air, but when the air was drawn out by means of an air pump, which Von Guericke had invented, an enormous force was required to separate them. This was explained by the pressure of the atmosphere on the outside being no longer balanced by an equal pressure inside.

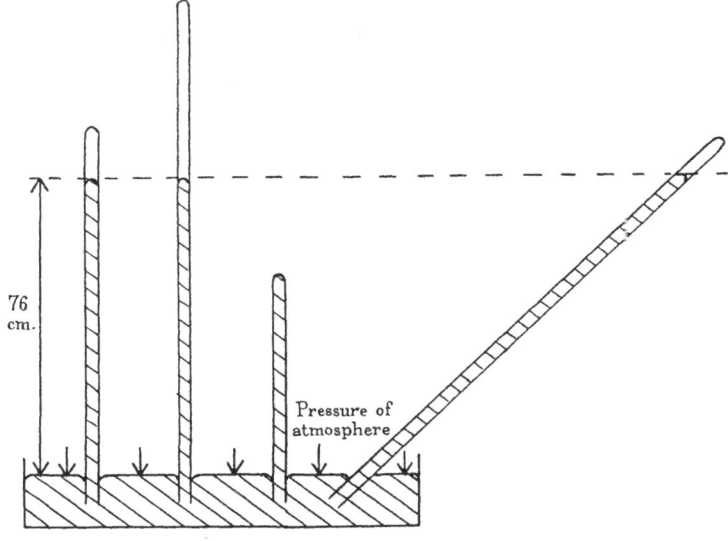

76 cm.

Pressure of atmosphere

Fig. 52

The experiment was repeated with hemispheres about a foot in diameter before the Emperor Ferdinand III in 1651, when two teams of sixteen horses were required to pull them apart.

About eleven years earlier, unknown to von Guericke, a vacuum was obtained with comparative ease by an Italian, Torricelli. Torricelli repeated the experiment with the long glass tube (described earlier), but he used, instead of water, the extremely heavy liquid, mercury (density 13·6 gm. per c.c.).

He filled a long glass tube, sealed at one end, with mercury and inverted it over mercury in a trough. The mercury did not remain in the tube, completely filling it, as the water had done but ran down a certain distance. Since there was nothing above it Torricelli had created a vacuum.

He found that whatever length of tube he used, the mercury would only stand at a vertical height of about 76 cm. If the tube was tilted from the vertical position, the mercury would run up it until its vertical height was 76 cm. (see Fig. 52).

Thus Nature's horror of a vacuum was not strong enough to keep a greater height than 76 cm. of mercury from falling. Torricelli put forward an entirely different theory. He suggested that the atmosphere was exerting a pressure on the surface of the mercury in the trough, and that the mercury falls in the tube until its weight is exactly balanced by this pressure. This theory was supported by the fact that the length of the mercury column varied slightly from day to day as the atmospheric pressure altered. Many men of his time, however, refused to believe that so great a weight could be held up by so light a substance as air.

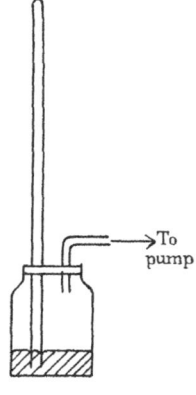

Fig. 53

A Frenchman, Pascal, took Torricelli's tube to the top of a church steeple in Paris. He reasoned that if the mercury column really is supported by the pressure of the atmosphere, the column should be shorter at a considerable height above the ground where there will be less air pressing down on the mercury in the trough. He could, however, detect no sinking of the mercury.

He next wrote to his brother-in-law who lived near the high mountains of the Auvergne, and asked him to take the apparatus to the top of the Puy de Dôme, which is 4800 ft. in height. This was done, and the mercury sank more than 6 cm. Thus the mercury column must be supported, at least in part, by the pressure of the atmosphere.

With the aid of an air pump, an experiment can be performed which clears up the point conclusively. Let the mercury trough take the form of an air-tight bottle (see Fig. 53), and pump out all the air from the bottle. If Torricelli's theory is true, the mercury in the tube should sink completely, and rise again when air is readmitted into the bottle. This is found to be the case.

The air pump.

An air pump has been used in foregoing experiments for creating a vacuum. Here is an explanation of how an air pump works.

It consists, in its simplest form (see Fig. 54), of a cylinder and a piston. At the base of the cylinder is a side tube, with a valve which will open only inwards, and the piston is fitted with a valve which opens only upwards. The side tube is connected by thick rubber tubing to the vessel to be evacuated, and the piston is worked up and down the cylinder, usually by a wheel and crank mechanism (see Fig. 46).

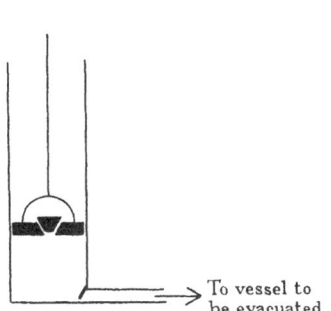

Fig. 54. Von Guericke's
air pump.

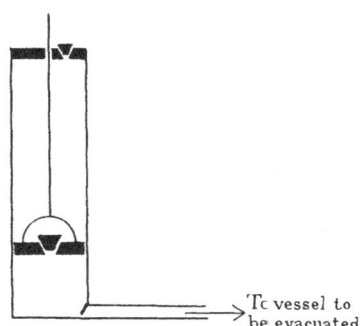

Fig. 55. Smeaton's improved
air pump.

To vessel to
be evacuated

To vessel to
be evacuated

When the piston is at the bottom of the cylinder all the air originally in the cylinder has been forced out through the valve in the piston. On raising the piston, a partial vacuum is left in the cylinder and air from the vessel being evacuated forces its way through the valve in the side tube into the cylinder. This air is expelled through the valve of the piston on the next downward stroke as it cannot return through the one-way valve in the side tube. The process continues until the vessel is practically evacuated.

The disadvantage of a pump of this very simple form, which was invented by von Guericke, is that the piston would be extremely hard to pull up after a few strokes, when the pressure of the atmosphere on the top is so much greater than the pressure of the air in the cylinder. Modern pumps, therefore, have a closed cylinder with a valve at the top opening only upwards (see Fig. 55). The air is thus forced right out of the cylinder on each upward stroke, and cannot re-enter owing to the valve at the top. This improvement was introduced by Smeaton.

There is another kind of air pump, which is used for producing compressed air. The simplest type is the ordinary bicycle pump. Draw a diagram similar to Fig. 54 but make the valves open the opposite way. Will this act as a compression pump?

Railway brakes.

The pressure of the atmosphere is used to operate railway brakes of the vacuum type. A train weighing hundreds of tons and travelling at 60 m.p.h. requires very efficient brakes.

Under each carriage there is a cylinder evacuated of air, fitted with a piston which is coupled to the brakes. By moving a lever in the engine or the guard's van, the atmosphere is admitted into the cylinder which forces up the piston and applies the brakes.

You have probably noticed, when two carriages are coupled, that flexible tubes are also locked together. Through these tubes steam from the engine passes, and by its motion can be made to suck out the air from the cylinders, drawing down the piston, and causing the brakes to be taken off again.

The principle of exhausting a vessel by means of a swiftly moving current of gas or vapour may be illustrated by blowing over the top of an open glass tube dipping in a tumbler of water. The water in the tube rises owing to some of the air being dragged out of the tube by the current of air passing above it.

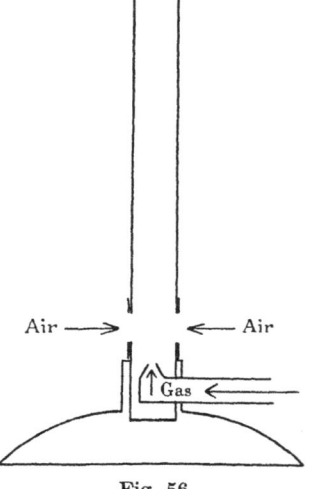

Fig. 56

When a strong gale breaks a shop window the glass usually falls outwards and not inwards as one might expect, owing to the "suction power" of a rapidly moving current of air.

Again, in a bunsen burner, the stream of gas issuing through a small nozzle (see Fig. 56) causes air to be sucked in through the side inlets.

Modern air pumps working on this principle can create an almost perfect vacuum, and are "over a million times more effective ' than the type of air pump described earlier in this chapter. When the apparatus has been exhausted as far as is possible by an ordinary pump the vapour from boiling mercury is made to sweep out the remaining air.

Barometers.

1. *The simple mercury barometer.* A barometer is an instrument for measuring the pressure of the atmosphere. The simplest type is a Torricelli tube filled carefully so that there are no air bubbles in the mercury, and upturned over a trough of mercury. The width of the tube makes no difference to the height of the mercury columns, since the pressure of the air, pressing on and transmitted through the mercury in the trough, has a greater area to push upon in the case of the wider tube, and can therefore support a greater weight of mercury. This demonstrates that the height of the mercury column (known as the barometric height) is a measure of pressure or force per unit area.

Pressure should really be measured in such units as lb. per sq. in., but it is convenient to express it merely as the length of the mercury column. In the same way one might buy one's sausages by the yard, and not bother to weigh them; there is, however, the important difference that the weight of a string of sausages depends on their cross-section, whereas the reading of the pressure of the atmosphere does not depend on the cross-section of the barometer tube.

The average barometric height is 76 cm. of mercury or about 30 in.

Now 1 cu. in. of mercury weighs about $\frac{1}{2}$ lb. Thus a column of mercury 30 in. high and 1 sq. in. in cross-section will have a volume of 30 cu. in., and hence a weight of $30 \times \frac{1}{2} = 15$ lb. The pressure represented by a column of mercury 30 in. high is therefore equivalent to a pressure of 15 lb. per sq. in.

Otto von Guericke had a barometer of this kind filled with water. Since water is so much lighter than mercury, the tube had to be unusually long.

Specific gravity of mercury $= 13 \cdot 6$.
Height of mercury barometer $= 30$ in.
$\therefore$ Height of water barometer $= 30 \times 13 \cdot 6$ in.
$= 34$ ft.

The top of von Guericke's barometer projected through a hole in the roof. Floating on the top of the water surface was a little wooden man. Now if the height of the mercury barometer rises or falls an inch, the height of the water barometer varies by more

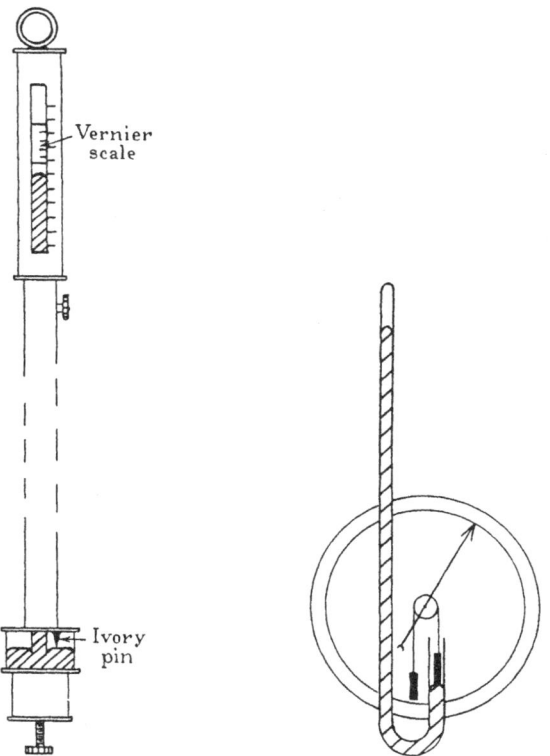

Fig. 57. Fortin barometer.   Fig. 58. Siphon barometer.

than 1 ft. Consequently the wooden man popped up above the roof when the pressure was high, and the weather fine. But when the pressure fell, and the weather was wet, he dropped out of sight below. Although quite a natural phenomenon, we are told that it occasioned strange rumours among the neighbours.

2. *The Fortin barometer.* It is convenient and usual to fix a barometer to a wall. If a simple mercury barometer and scale are thus fixed, there is a serious source of error due to the mercury level in the trough changing as the level in the tube changes. For the barometric height must be measured from the level in the trough.

The error might be overcome by moving the scale, but it is more convenient to have a fixed scale, and arrange for the mercury in the trough always to be at the same level. Thus the trough of the most common accurate barometer, known as the Fortin barometer (see Fig. 57), is made in the form of a leather bag which can be screwed up and down. There is an ivory pin fixed to the case of the instrument, and before any reading is taken, the surface of the mercury in the leather bag is made to touch the tip of this pin. Ivory is used in preference to metal because the latter is apt to become coated with a thin film of mercury.

The scale is usually graduated in both centimetres and inches, and there is a moving vernier enabling the reading to be taken accurately to two decimal places.

All laboratories possess an instrument of this kind, and you should certainly look at one and read it.

3. *The siphon barometer.* A type of barometer to be found in many houses is shown in Fig. 58. It consists of a glass tube containing mercury which is bent round at its lower end. An iron weight partially floats in the mercury, and is partially supported by a smaller iron weight hanging over a pulley carrying a pointer. When the atmospheric pressure increases, the mercury in the open end of the tube is forced down, the partially floating weight drops a little, and the pointer turns. The instrument is not very accurate, and the pulley is apt to stick.

4. *The aneroid barometer.* The aneroid barometer works on an entirely different principle from the mercury barometer. The word "aneroid" means "without liquid".

It consists of a partially evacuated box *B*, made of thin springy metal, which is corrugated in concentric circles, to cause it to cave in at the centre (see Fig. 59). It is prevented from collapsing altogether by a spring. When the atmospheric pressure increases, the box is pushed in slightly more than usual at the centre, and this movement is magnified by a system of levers which cause a pointer to move over a circular scale. When the atmospheric pressure decreases, the spring pulls the centre of the box out a little and the pointer moves the other way.

An aneroid barometer has to be graduated by comparing it with a Fortin barometer. It is not very accurate, and its reading

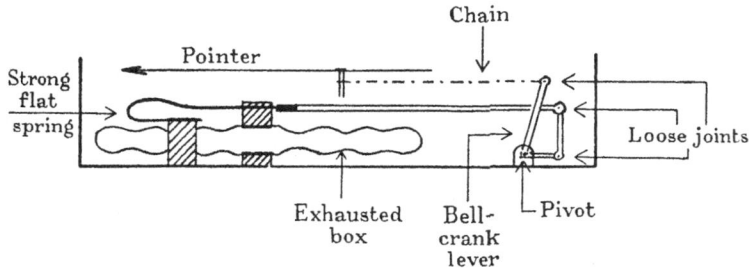

Fig. 59. Aneroid barometer.

should be tested at intervals. It has the great advantage of being compact and portable, and can be bought in a size no larger than that of a watch.

The barograph. The barograph is a form of aneroid barometer which automatically records the atmospheric pressure on a slowly rotating drum, driven by clockwork. The paper round the drum is replaced once a week, during which time the drum makes one complete revolution.

The height of the atmosphere.

We can calculate the height of the atmosphere by using two quantities determined in the laboratory—the density of air and the density of mercury. We know that the pressure of the atmosphere, which is due simply to its weight, can support a column of mercury 76 cm. high. Let us therefore imagine two columns of mercury and air, each with a cross-section of 1 sq. cm., that of mercury being 76 cm. long, and that of air being as high as the atmosphere. These two columns must be of equal weight.

Let   Height of mercury column $= 76$ cm.

         Height of air column $= h$ cm.

         Cross-section of each column $= 1$ sq. cm.

         Density of mercury $= 13 \cdot 6$ gm. per c.c.

Density of air $= 0 \cdot 0012$ gm. per c.c.
Volume of mercury column $= 76 \times 1$ c.c.
$\therefore$ Weight of mercury column $= 76 \times 13 \cdot 6$ gm.
Similarly   Weight of air column $= h \times 1 \times 0 \cdot 0012$ gm.

But these columns have equal weight,

$$\therefore \quad 0 \cdot 0012h = 76 \times 13 \cdot 6,$$

$$\therefore \qquad h = \frac{76 \times 13 \cdot 6}{0 \cdot 0012}$$

$$= 861,200 \text{ cm.}$$

$$= 8 \cdot 612 \text{ kilometres}$$

$$= \text{approx. } 5\tfrac{3}{8} \text{ miles.}$$

(1 kilometre $=$ approx. $\tfrac{5}{8}$ mile.)

This calculation gives us the height of the atmosphere as 5 miles. But we have definite evidence that the atmosphere extends very much higher than this. Climbers, for instance, have ascended nearly to the summit of Mount Everest, which is more than 5 miles high, and there is still just enough air for them to breathe, the pressure being about one-quarter of normal atmospheric pressure. Aeroplane ascents of $8\tfrac{1}{2}$ miles, and balloon ascents of $10\tfrac{1}{2}$ miles (by Monsieur Piccard) and $13\tfrac{1}{2}$ miles (by a Soviet balloon), have been made. Both aeroplanes and balloons, of course, require air to enable them to fly.

There is further evidence that the atmosphere must extend to a height of at least 100 miles. Most people have seen, on a dark clear night, what are known popularly as shooting stars. These look like stars, suddenly appearing, shooting with great speed across the sky, and being as suddenly extinguished. Actually they are not stars at all, but quite small bodies known as meteors. There are millions of these small bodies wandering through space, absolutely cold and non-luminous. When any of them enter the earth's atmosphere, owing to their great speed they become white hot by friction of the air. Most meteorites are burnt up completely—which is fortunate, as otherwise we should be subjected to a terrible bombardment. But sometimes, bodies weighing several pounds reach the earth. A huge meteorite fell in Siberia in 1904, flattening all the surrounding trees by the great blast of hot air which it caused, and in certain parts of the earth are craters, up to hundreds of feet across, almost certainly due to meteorites.

However, the point which concerns us here is that meteorites first light up at a height of about 100 miles, proving that the atmosphere must extend at least as far as that.

How then are we to reconcile this with our calculation? The answer is to be found in the compressibility of air. Air at sea level has a great weight of air above it, and is therefore compressed. At a greater altitude there is less air overhead pressing down, and thus the air is less compressed. As we ascend a mountain we find that the pressure of the air steadily decreases.

Compressed air has a greater density than air which is less compressed. In our calculation we assumed the density of the whole atmosphere to be the same as that of the most compressed air at sea level. Since the density of air decreases with height, it is clear that the atmosphere must be considerably higher, in order to support 76 cm. of mercury, than if its density remained constant all the way up.

It is an interesting experiment to carry a pocket aneroid barometer while climbing a mountain. There is a fall of pressure of about 1 inch (the aneroid barometers are graduated in inches of mercury by comparison with a Fortin barometer) for every 1000 ft. ascended. Thus on climbing Helvellyn, in the English Lake District, which is 3119 ft. high, there is a fall of about 3 in. in the pressure. It will be remembered that this experiment was first performed with a mercury barometer, at Pascal's suggestion, in the mountains of the Auvergne.

One of the chief difficulties which faced the climbers of Mount Everest was the difficulty of breathing. Mount Everest is 29,000 ft., i.e. about $5\frac{1}{2}$ miles high. At this height, the pressure of the atmosphere is only about one-quarter of its value at sea level. It is therefore necessary to take four breaths in order to obtain the same amount of oxygen as is obtained in one breath at sea level, and in the last stages of the climb the party have to pause between each step, and breathe several times.

There is an interesting difference of opinion between climbers as to whether it is better to carry oxygen cylinders, or to acclimatise oneself slowly to the gradually decreasing pressure of oxygen, and climb unencumbered with heavy cylinders.

The decrease in the pressure of the atmosphere with height is utilised in the altimeter, the instrument by means of which aeroplanes and airships can tell their height. Most altimeters consist of a form of aneroid barometer.

Weather forecasting.

The barometer is often called the weather glass, and is marked with such words as Fine, Change, Wet, for from a reading of the atmospheric pressure and the way it is changing, we can obtain an indication of what the weather will be like. If the pressure is

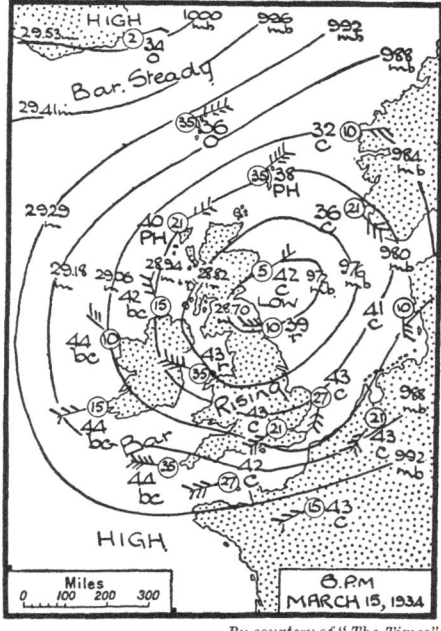

By courtesy of " The Times"

Fig. 60. A cyclone or depression.

slowly and steadily rising, we may expect fine, settled weather; and if it is falling, wet, unsettled weather.

Every day weather forecasts issued by the Meteorological Office in London are published in the newspapers and broadcast by wireless. These forecasts are based upon readings of the barometer taken at some 300 stations situated all over Europe. At 7 a.m., 1 p.m. and 6 p.m. the observations are simultaneously made at all stations and sent by telephone, telegraph, or wireless

to some central collecting office. Paris collects all reports from
South-West Europe, Berlin from the North-East, London from
the British Isles and ships out on the Atlantic.

The barometer readings are plotted on a map by means of
lines called *isobars*, which connect places having the same atmospheric pressure.

Fig. 60 shows one of these maps. The pressure is marked in
inches, and also in a special unit, the millibar, mb. (1000 mb.

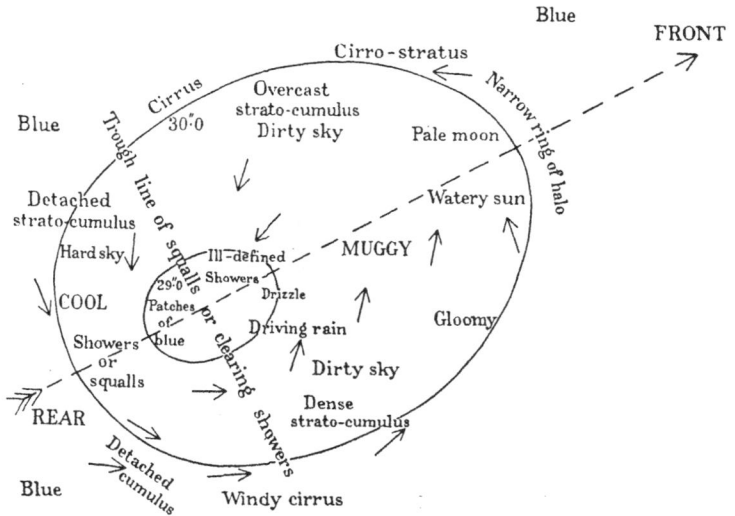

Fig. 61. The weather experienced in a typical depression or cyclone.

=29·53 in.), and it will be seen that the isobars are drawn at
differences of pressure of 4 millibars.

If hundreds of these maps are examined, they are found to
possess striking similarities. The isobars are usually curves, surrounding some central area where the pressure may be high or
low. In Fig. 60 the pressure at the centre of the system is low.
This is known as a *region of low pressure*, a *depression*, or a
cyclone. Such a system moves about as a whole. Depressions
breed over the Atlantic, and usually approach the British Isles
from the South-West or West. While they are passing over, wet

unsettled weather is experienced. Fig. 61 shows the kind of
weather usually experienced in a typical cyclone; hence, when
a cyclone is approaching, the sequence of weather can be forecast.

When the pressure in the centre is high, the system is called a
region of high pressure, or an *anti-cyclone*. These systems are

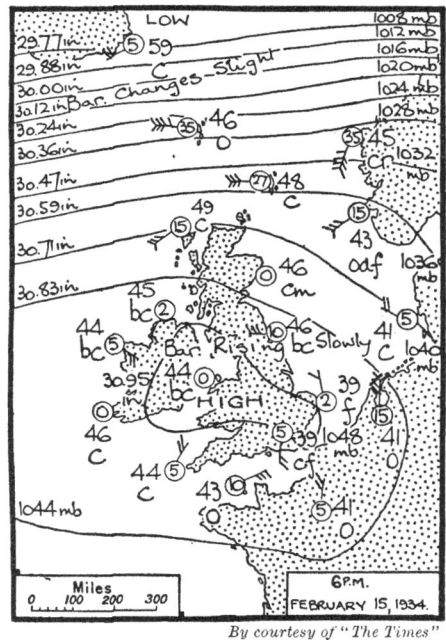

By courtesy of "The Times"

Fig. 62. An anti-cyclone.

usually fairly stationary, and are associated with normal fine
weather.

The wireless announcer, when reading the familiar forecast
each evening, often makes use of other terms besides cyclone and
anti-cyclone, such as associated secondary, ridge and wedge of
high pressure, trough of low pressure, and **V**-shaped depression.
These other types of pressure distribution are modifications of
cyclones and anti-cyclones. Sometimes a cyclone breaks up into

two parts with two regions of low pressure, and the new region of low pressure is known as a secondary depression. Ridges and wedges of high pressure are regions of high pressure between two cyclones, while troughs and V-shaped depressions are portions of cyclones.

Besides readings of the atmospheric pressure, observations are also taken at meteorological stations of temperature and wind velocity. The small figures on the map indicate temperature in degrees Fahrenheit, and the larger figures in a circle the wind velocity in miles per hour. The number of feathers on the tail of the arrows also show the force of the wind. Thus 2 feathers indicate a light breeze, 6 a strong wind, and 12 a hurricane.

It will be noticed that the winds are blowing round the region of low pressure in an anti-clockwise direction (opposite to the direction of the hands of a clock). In the Northern Hemisphere, winds always circle in this direction round a region of low pressure, and in the opposite direction, clockwise, round an anti-cyclone.

It might be expected that winds would blow direct from a region of high pressure to regions of lower pressure, just as water flows downhill. We see by looking at the maps that this does happen to a certain extent, for the winds do blow across the isobars. Furthermore, the winds are strongest where the isobars are closest together, since here there is a more sudden change from high to low pressure. However, the direction of the wind is modified by the rotation of the earth, causing a rotary motion round cyclones and anti-cyclones.

The letters on the map stand for the weather conditions: *b* for blue sky, *c* for cloudy, *o* for overcast, *r* for rain.

Since weather forecasting depends on the collection of data from many widely distant stations, it only became possible with the invention of electric telegraphy, and was first attempted in 1860 by Admiral Fitzroy. The wind velocity scale represented by the feathers on the tail of arrows was invented by Admiral Beaufort in 1805, and is known as the Beaufort Scale. The original descriptions of the strength of the wind had reference to the navigation of a naval sailing vessel.

You should certainly look at the weather map which is published every day in *The Times*, *The Morning Post*, and *The Daily Telegraph*. During the Great War, no newspaper published weather maps, as such information had considerable military value. For instance, the Germans lost in violent storms several airships, which they would certainly have warned and recalled in time had they possessed information known to the British weather experts.

An experiment on the compressibility of air.

To find how the volume of air depends on its pressure. When calculating the height of the atmosphere a wrong result was obtained because no allowance was made for the compressibility of air.

In the following experiment a fixed quantity of air is compressed, and the change in volume noted. Fig. 63 shows a convenient form of apparatus for the purpose. The air under test is contained in a glass tube graduated in cubic inches, and having at its upper end a pressure gauge graduated in lb. per sq. in. The lower end of the tube is bent round in the shape of a letter U and contains water. The water prevents the air escaping, and enables it to be subjected to different pressures. *V* is an ordinary bicycle tyre valve through which air can be pumped. In this way the water is forced down on the one side, and up on the other, compressing the air under test. (N.B. The air pumped in does not mix with the air under test, which remains the same fixed quantity throughout.)

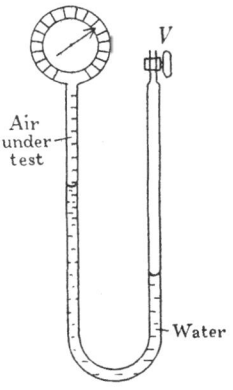

Fig. 63. Apparatus for verifying Boyle's Law.

Draw two columns on paper, and head them Pressure and Volume. Unscrew the valve so that the pressure of the air under test is the same as atmospheric pressure. Read the pressure gauge. To obtain the volume of the air at atmospheric pressure, read the level of the meniscus of water. Now screw up the valve, pump a little, and again read the pressure and volume. Repeat, obtaining in all six pairs of readings.

Examine your readings, and see if there is any simple relation between them. If not, read the volume of the air under test when the pressure is 15 lb. per sq. in., and again at exactly double this pressure, i.e. 30 lb. per sq. in. You will find that when the pressure is doubled, the volume is almost exactly halved. Here is a clue which may lead to a simple law.

Suppose that there is a law connecting the pressure and volume, so that trebling the pressure causes the volume to become one-third of its original value, and so on. How is it possible to discover whether the six pairs of readings taken conform to such a law?

Let us imagine for simplicity that at a pressure of 15 lb. per sq. in. the volume is exactly 4 cu. in.

Assuming our supposed law to be true,

At pressure of 30 lb. per sq. in., volume is 2 cu. in.
,, ,, 60 ,, ,, ,, ,, 1 ,,
,, ,, 7½ ,, ,, ,, ,, 8 ,,

and so on.

Now draw a third column and head it "Pressure × Volume". We obtain the following:

Pressure, p (lb. per sq. in.)	Volume, v (cu. in.)	pv
15	4	60
30	2	60
60	1	60
7½	8	60
5	12	60

It will be seen that the values in the third column are all the same.

The product pv should be worked out for all the readings actually taken. If they are all the same (not necessarily 60), the law that doubling the pressure means halving the volume, etc. will have been proved to be generally true.

The first man to discover this law was Robert Boyle (1627–91). His discovery may be summed up neatly in a statement known as **Boyle's Law:**

The pressure and volume of a fixed quantity of a gas are inversely proportional, provided that the temperature is kept constant.

Examine carefully this statement of the law. First of all, instead of air, the word "gas" is used. The law holds for all other gases besides air. This can be proved by filling the apparatus with other gases, such as hydrogen, instead of air. The word "inversely" means "upside down"—if anything is inverted it is turned up-side down.

To state the law in mathematical symbols:

$$p \propto \frac{1}{v},$$

$$\therefore \ pv = \text{constant}.$$

Thus if the pressure is made three times as great, the volume becomes (turning three upside down, as it were) one-third as great.

The temperature of the gas must remain constant, as raising its temperature has the effect of increasing its volume or pressure or both. Motor tyres left exposed to hot sunlight have been known to burst, due to the increase in pressure resulting from the heat.

No special precautions were taken during the experiment to keep the temperature constant, but unless the apparatus was exposed to strong sunlight, it can be assumed that the temperature remained steady.

Scientific laws.

The word "law" is a term which has a special meaning when used in the scientific sense. The State "lays down the law" that no person shall ride a bicycle or other vehicle after dark without lights. Such a law is passed by Parliament, and any person caught disobeying it is punished.

Scientific laws, on the other hand, are not "passed"; nor is there any element of compulsion about them. Gases do not disobey Boyle's Law at their peril; for when, at high pressures, they cease to obey it accurately, it is the law which has to be corrected.

A scientific law is simply a general statement which summarises many facts. It does not attempt an explanation like a theory, but is merely a summary. Boyle's Law is a general statement of the behaviour of gases when they are compressed at constant temperature.

Storing gases under pressure.

When one orders from the manufacturer a quantity of oxygen or hydrogen, a lorry does not arrive towing a balloon containing the gas. It is sent compressed, at a pressure of perhaps 120 atmospheres, in a steel cylinder. Applying Boyle's Law, if the pressure is 120 times ordinary atmospheric pressure, when the tap of the cylinder is opened, the gas will escape, and expand to 120 times the internal volume of the cylinder.

Application of Boyle's Law.

Determination of the pressure of the water supply. This experiment, to determine the pressure of the water supply, is a simple application of Boyle's Law.

Take a glass tube sealed at one end (between 40 cm. and 50 cm. is a convenient length), introduce a column of water a few cm. long, and then fit the open end with a rubber bung through which passes a piece of glass tubing. Connect the glass tube to the tap with pressure tubing.

Clamp the glass tube firmly in a vertical position (see Fig. 64) and turn on the tap; measure the length of the column of air, which will have been considerably compressed. (The tube should be of thick glass and strongly sealed.)

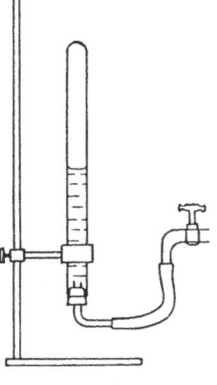

It is now necessary to find the length of the air column when the air is at atmospheric pressure. Remove the rubber bung under the surface of water in a suitable vessel and adjust

Fig. 64

the position of the glass tube so that the level of the water inside it is the same as the level outside. The pressure of the air is now atmospheric; find the length of the air column. Read the barometer.

The result is worked out as follows:

Length of air column under pressure of water supply = 21·7 cm.
Length of air column under atmospheric pressure   = 39·6 cm.
Pressure of atmosphere   = 74·6 cm. of mercury.
Let pressure of water supply = p cm. of mercury.
By Boyle's Law, pv = constant.

We can represent the volumes of the air by the lengths of the air columns if the tube is of uniform cross-section.

$$\therefore \quad p \times 21{\cdot}7 = 74{\cdot}6 \times 39{\cdot}6,$$
$$p = \frac{74{\cdot}6 \times 39{\cdot}6}{21{\cdot}7}$$
$$= 136 \text{ cm. of mercury.}$$

Since   76 cm. of mercury ≡ 14·7 lb. per sq. in.,

$$p = 136 \times \frac{14 \cdot 7}{76} \text{ lb. per sq. in.}$$
$$= 26 \cdot 3 \text{ lb. per sq. in.}$$

A similar method is used for finding the pressure at the bottom of the sea, and hence the depth. A glass tube, kept vertical by a heavy sinker, is let down and the distance the sea water enters the tube is indicated by the discoloration of silver chromate with which the interior of the tube is coated. This apparatus, invented by Lord Kelvin, replaced the old line and sinker method, which is unsuitable for deep sea sounding.

Theory of the air pump.

When an air pump (see Fig. 54) is used to exhaust a vessel, it draws out a smaller weight of air at each successive stroke. Although the same volume of air is removed at each stroke (equal to the volume of the barrel), the pressure of this air becomes less and less.

Calculation of the pressure in the exhausted vessel, called the receiver, after n strokes.

Let V = volume of receiver,
                           v = volume of barrel of pump.

On the first stroke, a volume V of air expands to $V + v$.

If p_1 = pressure of volume V,
                    p_2 = pressure of volume $V + v$,

by Boyle's Law,

$$p_1 V = p_2 (V + v),$$
$$p_2 = \frac{V}{V + v} \cdot p_1.$$

After two strokes the pressure will be $\left(\dfrac{V}{V + v}\right)^2 p_1$.

Thus, after n strokes, the pressure in the receiver will be $\left(\dfrac{V}{V + v}\right)^n$ of its original value.

Water pumps.

The syringe. The syringe which is used for spraying roses or fruit trees consists of a cylinder fitted with a piston (see Fig. 65). When the nozzle is placed under a liquid and the piston is drawn up, a partial vacuum is left in the cylinder. The atmosphere pressing on the surface of the liquid forces it into the cylinder.

The lift pump. The lift pump is the type of pump one sometimes sees in a country cottage which has to depend on a well for its water supply. It consists (see Fig. 66) of a cylinder or barrel having a valve at the bottom, known as the foot valve, which opens upwards only, and a piston containing a valve which also opens upwards. When the piston is raised the pressure of the atmosphere keeps the piston valve closed, and forces the water from the well up through the foot valve into the partial vacuum

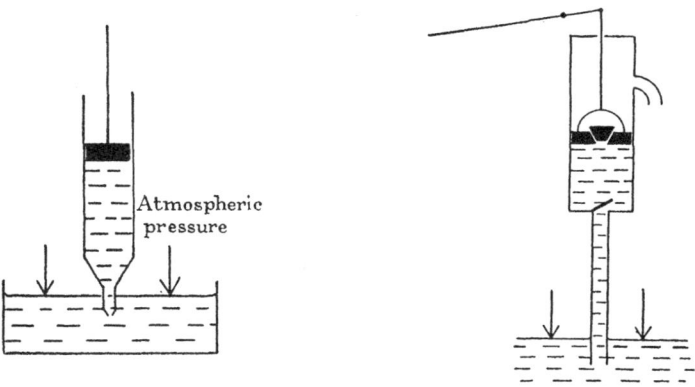

Fig. 65. Syringe.   Fig. 66. Lift pump.

left in the barrel. On lowering the piston the water passes through the piston valve and on the next upstroke it is lifted on the piston, and issues from a spout at the top. At the same time more water is forced through the foot valve by the atmosphere. The fact that the water is lifted on the piston gives the pump its name. The water comes out of the spout when the handle is pushed down, for this pulls up the piston.

Since the atmosphere can support a column of water only 34 ft. high, the foot valve must be less than 34 ft. above the level of the water in the well. In practice, mainly owing to leakage between the piston and the cylinder, the vacuum in the barrel is by no means perfect, and the water will not rise much more than 25 ft.

Lift pumps are sometimes used for pumping water out of deep wells, and then the barrel of the pump is placed down in the well shaft. The piston rod is long and can be operated from the top of the well. On each downstroke of the piston a barrel-full of water is forced above the piston valve until eventually the water comes out of the spout at the top of the well.

The force pump. The force pump is used for forcing water to a

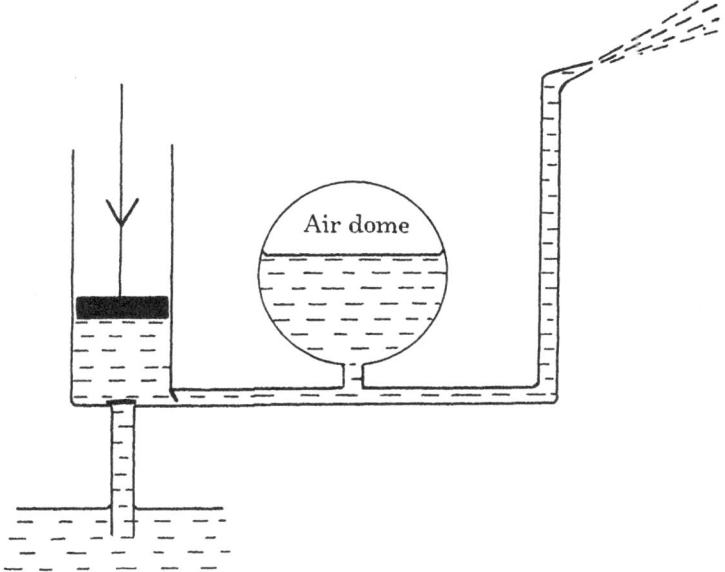

Fig. 67. Force pump.

considerable height and, in a modified form, is the type of pump to be found in the old type of fire engines.

It consists (see Fig. 67) of a barrel with a foot valve opening upwards only, a solid piston without a valve, and a side tube at the bottom of the barrel opening outwards only. When the piston is raised, a partial vacuum is left in the barrel, and the pressure of the atmosphere forces water up through the foot valve so long as the foot valve is not higher than 34 ft. (theoretical height) above the level of the water in the reservoir. On pushing the piston down the water is forced through the side valve into the side tube, the pressure of the water closing the foot valve. At each downstroke of the piston a barrel-full of water is forced into the side tube, and the latter may be as high as is consistent with the force available to push the piston down. An air dome is often used with the pump. The air in the dome becomes compressed and forces the water out of the spout in a fairly continuous stream. Otherwise the water is ejected in spurts at each downstroke of the piston.

It is interesting to note that the piston rod of the force pump must be short and thick, to prevent it buckling, since the maximum force is the downstroke. In the lift pump, however, the piston rod may be long and thin, since most of the work is done on the upstroke, when it is in tension, and there is no tendency to buckle.

Rotary pumps. The most powerful pumps, used in modern fire engines, and for pumping water out of mines and docks, work on a different principle from those described above. They consist essentially of a wheel with curved blades which is rotated at high speed, rather like an electric fan or a propeller. When an electric fan revolves it causes a draught, drawing air from behind, and pushing it in front. The strong suction power of an electric fan is the essential feature in most electric vacuum cleaners, which are, in fact, rotary air pumps.

It is the forcing of water backwards by the propeller of a ship which causes her to move forwards. Water can be raised to a considerable height by a rotary pump, which has the great advantage of containing no valves to stick or become choked. The latest fire engines can throw a jet as high as 340 ft., higher than the spire of most churches.

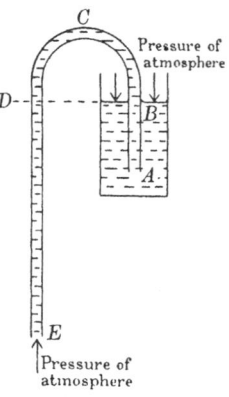

Fig. 68. The siphon.

The siphon.

It is often convenient to empty a deep and narrow vessel of water by harnessing the atmospheric pressure. The only apparatus required is a piece of bent glass tubing. Completely fill the tube with water, then place one end beneath the surface of the water in the vessel, and let the other end hang below the base (see Fig. 68). Water runs out of the tube so long as the vertical height between B and C is not greater than 34 ft. (since the atmosphere cannot support a greater height of water than this).

It is essential that E shall be lower than the surface of the water as the vessel is being emptied. For the pressure of the atmosphere on the surface of the water can only overcome the equal and opposite pressure of the atmosphere on the open end of the outside arm of the tube if it is helped by a greater weight of water in the outside arm than it has to force up the inside arm.

Pressure at D = pressure at B = pressure of the atmosphere.

∴ Downward pressure of water at E = pressure of the atmosphere + pressure due to column of water DE.

In the soda water siphon, the pressure of the gas given off by the soda water is greater than the pressure of the outside atmosphere, and consequently the spout of the siphon can be higher than the surface of the liquid.

Summary

Owing to the great weight of air above the earth, **the atmosphere exerts a pressure of about 15 lb. per sq. in. at sea level. This pressure will support a column of mercury approximately 76 cm. high.** With increasing height above sea level, the atmospheric pressure decreases.

There are two types of barometer, the mercury barometer, of which Fortin's is the most accurate, and the aneroid barometer. Weather forecasting is done by drawing on a map lines called isobars, which join places at which the atmospheric pressure is the same. Isobars usually curve round a region of high pressure known as an anti-cyclone, or one of low pressure known as a cyclone. Anti-cyclones are associated with fine weather, and cyclones with wet and stormy weather.

Boyle's Law.

The pressure and volume of a fixed quantity of a gas are inversely proportional, provided that the temperature is kept constant.

The syringe, lift pump, force pump and siphon depend for their action on the pressure of the atmosphere.

QUESTIONS

1. (*a*) Describe how you would set up and use a simple mercury barometer.

(*b*) Why is water an unsuitable liquid for use in a barometer?

(*c*) Why does the diameter of the barometer tube make no difference to the readings?

(*d*) Is the reading affected if the barometer tube is inclined to the vertical?

(*e*) Explain, with the aid of a diagram, why more accurate readings are possible with a Fortin barometer.

2. How would you find out whether there was any air at the top of a barometer tube without the aid of another barometer? What effect would such air have on the reading?

How will the size of the error of the barometer's reading be affected by taking it (*a*) up a mountain, (*b*) down a coal mine?

3. Explain carefully:

(*a*) The bung must be removed from a barrel in order that the contained liquid may flow out when the tap at the bottom is opened.

(*b*) A leather sucker, i.e. a piece of damp leather on the end of a string, can be used to lift a stone.

(*c*) A soap bubble does not collapse although the atmosphere is exerting a force on it of 20 or 30 lb.

(*d*) Will a barometer read lower in a building than in the open, owing to the roof?

4. Describe, with a diagram, how the self-filling apparatus of a fountain pen works. (If you have an old one, pull out the inside and look at it.)

5. (*a*) Explain how it is possible to suck up lemonade through a straw.

(*b*) Why is the straw useless if it has a hole in the side?

(*c*) Is there any limit to the length of the tube which could be used?

6. The original Magdeburg hemispheres had a diameter of about 14 in. Taking atmospheric pressure as 15 lb. per sq. in., find the force required to separate them. (Calculate the force exerted by the atmosphere on a circle of diameter 14 in. Why?)

7. The height of a mercury barometer is 76 cm. What would be the height if oil of density 0·90 gm. per c.c. were used? (Density of mercury = 13·6 gm. per c.c.)

8. Invent an air pump for filling a metal cylinder, having a tap, with compressed air. Point out how your pump differs from an ordinary bicycle pump.

9. Fig. 69 shows an automatic flushing tank. Explain how it works.

10. A vessel 100 cm. deep contains mercury to a depth of 90 cm. What is the greatest depth of mercury which can be siphoned out? What is the shortest possible length of the outer arm of the siphon for this purpose? Is there any advantage in further increasing its length?

If, while the siphon was working, the atmospheric pressure were removed for a few seconds and then restored, what would happen?

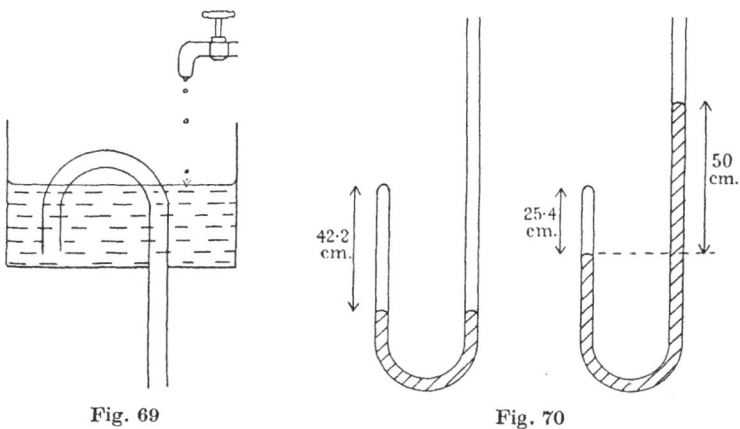

Fig. 69   Fig. 70

11. If the cyclone in Fig. 60 is moving South, describe the sequence of weather conditions to be expected in London. [Use the information in Fig. 61.]

12. The volume of the air in a motor tyre is 1200 cu. in. and its pressure is 32 lb. per sq. in. What volume would this air occupy at ordinary atmospheric pressure, i.e. 15 lb. per sq. in.?

13. Air is trapped in a U-tube, sealed at one end, by mercury (see Fig. 70). When the mercury levels in the two arms are the same, the length of the air column is 42·2 cm. When more mercury is poured in so that the difference in levels is 50 cm., the length of the air column is 25·4 cm. Find the pressure of the atmosphere.

14. Two closed vessels, A, of volume 50 cu. in., and B, of volume 20 cu. in., are connected by a narrow pipe (of negligible volume) with

a closed tap in it. Originally A contains air at a pressure of 30 lb. per sq. in., and B contains air at a pressure of 18 lb. per sq. in. The tap is opened. What is the new pressure inside the vessels?

School Certificate Questions

15. Explain how a barometer measures the pressure of the atmosphere.

A man climbing a mountain from sea level observes a drop of 4 cm. in the height of the barometer. Find the approximate height of the mountain, assuming the density of mercury to be 13·6 gm. per c.c. and that of air to be 1·29 gm. per litre.

16. Describe the common pump. Why is there a limit to the depth from which water can be raised by means of it? What kind of pump would be suitable for raising water from a well 25 ft. deep to a tank 50 ft. above the ground? Illustrate your answer by diagrams.

17. The diagram (Fig. 71) shows a petrol pump in which A is a ratchet arrangement, B the body of the pump and C the petrol delivery tap. Describe clearly how the pump works and how it delivers a fixed volume of petrol. What would be the effect on the working of the pump of blocking up the outlet D?

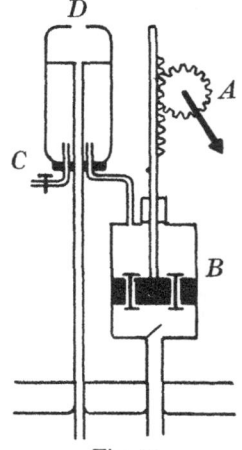

18. Explain clearly the action of a siphon. Water is flowing out of a vessel through a siphon. What will happen if a small crack develops (a) in the shorter limb of the siphon, (b) in the longer limb?

19. State Boyle's Law.

The receiver of an air pump has a volume of 2 litres and contains air at atmospheric pressure. The cylinder of the pump has a volume of 250 c.c. What will be the pressure of the air in the receiver after (a) one stroke, (b) three strokes of the pump?

20. State Boyle's Law for gases and describe briefly how you would verify it experimentally.

Fig. 71

A capillary tube, sealed at one end, contains air enclosed by a thread of mercury 10 cm. long. When the tube is horizontal, the column of air is 25 cm. long. What would you expect the length of the air column to be when the tube is held vertically with the open end upwards? (Height of barometer = 76 cm. of mercury.)

21. A bubble of gas released at the bottom of a lake increases to ten times its volume when it rises to the surface. How deep is the lake? The atmospheric pressure is 1033 gm. wt. per sq. cm.

22. State Boyle's Law.

A short cylinder, closed at one end, is sunk, open end downwards, to the bottom of a lake. On drawing it up again, it is found to have been wetted inside for two-thirds its height. What is the depth of the lake? The barometric height at the time is 75 cm. and the sp. gr. of mercury is 13·5.

23. A barometer which contains some air above the mercury reads 76 cm. when the correct height is 77 cm. If the height of the top of the barometer tube is 85 cm. above the level of the mercury in the cistern, find the true barometric height when the mercury reads 75 cm.

24. State Boyle's Law.

At a time when the barometric height was 75·7 the following observations were made with a Boyle's Law apparatus (similar to that in Fig. 70), the numbers representing heights above the table.

Mercury level in closed limb (cm.)	Mercury level in open limb (cm.)
46·6	15·9
57·2	41·6
63·8	63·1
67·7	80·4
72·2	106·5

Deduce in each case numbers (p and v) corresponding to the pressure and volume. Plot p against $1/v$. Draw the graph which best represents the observations. What shape would you expect this graph to be from Boyle's Law?

Chapter V

THE PRINCIPLE OF ARCHIMEDES

Buoyancy.

If a cork is pushed under the surface of water and released, it bobs up. The water is said to *buoy* up the cork.

We can investigate this phenomenon, known as *buoyancy*, by the following experiments. Suspend a metal cube by means of a piece of thread from a spring balance and find its weight in air. Now immerse the cube completely in water and read the balance again. The apparent weight of the cube will be found to have decreased considerably.

Repeat the experiment with cubes of different metals, but equal volume. The apparent loss in weight will be found to be the same in each case.

Now take a cube of twice the size; the apparent loss in weight will be found to be twice as great as it was before. The apparent loss in weight, therefore, depends on the volume of the cube, but not on the substance of which it is made.

To find whether changing the liquid makes any difference, weigh the cube in paraffin. The apparent loss in weight will be found to be less in paraffin than it was in water.

Readings from an experiment.

$$\begin{array}{lll}
\text{Weight of iron cube in air} & = 62 \text{ gm.} \\
\quad\quad ,, \quad\quad ,, \quad\quad ,, \quad\quad \text{water} & = 54 \text{ gm.} \\
\quad\quad ,, \quad\quad ,, \quad\quad ,, \quad\quad \text{paraffin} & = 55 \text{ gm.} \\
\therefore \text{ Apparent loss in weight in water} & = 8 \text{ gm.} \\
\text{and} \quad\quad ,, \quad\quad ,, \quad\quad ,, \quad\quad ,, \quad \text{paraffin} & = 7 \text{ gm.}
\end{array}$$

The apparent loss in weight is clearly due to a push-up or upthrust of the liquid, and we have shown that this upthrust depends on the volume of solid and the nature of the liquid, but not on the substance of which the solid is composed.

The Principles of Archimedes.

More than two thousand years ago, Archimedes discovered the following law, which is known as the **Principle of Archimedes. When a body is totally or partially immersed in a fluid it**

experiences an upthrust equal to the weight of fluid displaced.

The word *fluid* connotes gas as well as liquid. We shall consider the case of gases later.

Let us take a simple example to illustrate the meaning of the principle. Suppose a cube of volume 8 c.c. is immersed in water. It displaces 8 c.c. of water. Since 1 c.c. of water weighs 1 gm., the weight of water displaced is 8 gm. Archimedes' Principle tells us, therefore, that the upthrust is equal to the weight of water displaced, i.e. 8 gm.

Look at the readings just given of the weight of a cube in air and water. The apparent loss in weight of the cube, which is equal to the upthrust by the water, is 8 gm. Arguing backwards this time, we deduce that the weight of water displaced is 8 gm., and the volume of water displaced is 8 c.c. Hence the volume of the cube must have been 8 c.c.

Experimental proof of Archimedes' Principle.

Cylinder and bucket experiment. A straightforward though rough experimental proof of Archimedes' Principle may be performed with a solid cylinder which fits exactly into a hollow

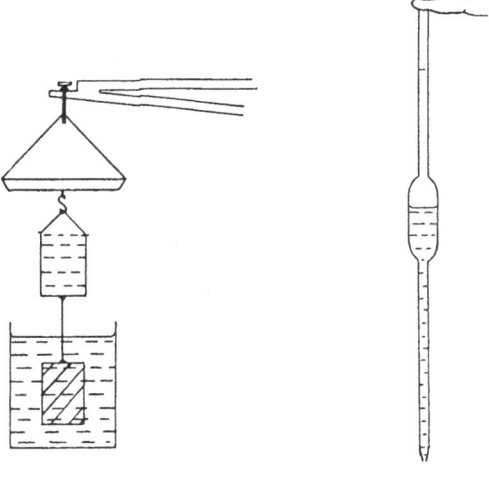

Fig. 72   Fig. 73. Pipette.

cylinder. Suspend the cylinders from one arm of a balance with the hollow cylinder uppermost, and the solid cylinder hanging below. Some balances are made with one pan specially short for experiments of this type (see Fig. 72). Place weights in the other pan until the cylinders, which are hanging in air, are exactly counterpoised. These weights need not be recorded but must not be altered during the rest of the experiment.

Now immerse the solid cylinder in water. Owing to the up-thrust of the water the counterpoising weights will be too heavy. Restore the balance by filling the hollow cylinder just up to the brim with water. It is convenient to do this by means of a pipette (an instrument which consists of a glass tube up which water may be sucked) (Fig. 73). If the finger is placed over the upper end, the rate at which the water runs out can be controlled, and the last few c.c. added drop by drop until a balance is obtained.

Thus the upthrust on the lower solid cylinder is counteracted by the weight of water in the hollow cylinder. Since the volume of the hollow cylinder is exactly equal to that of the solid cylinder, the upthrust must be equal to the weight of water displaced.

Why is this experiment only a rough one?

Why there is an upthrust.

Why is there an upthrust on a body when it is immersed in a liquid? One might expect that when it was under the surface there would be a downward instead of an upward force on the body, owing to the weight of liquid above it.

Imagine a cube suspended in a liquid as shown in Fig. 74. There is certainly a downward force on the top, represented by the arrows, but there is also an upward force on the bottom which is greater, since the pressure in a liquid increases with the depth. It must be remembered that the pressure in a liquid not only acts downwards but in all directions, and the pressure on the bottom of the cube can only act upwards.

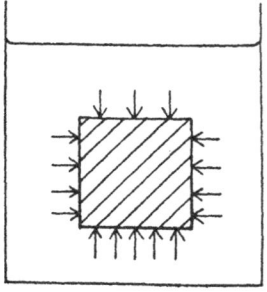

Fig. 74

The forces exerted by the liquid on the sides balance out.

Hence there is a resultant upthrust on the cube which is equal to the difference between the upward force on the bottom and the downward force on the top.

Simple theoretical proof of Archimedes' Principle.

A theoretical proof means a proof on paper from first principles, rather like the proof of a theorem in geometry. From our knowledge of pressure in liquids we shall prove that Archimedes' Principle must be true. Theoretical physics of this kind often leads to discoveries of great importance. Thus Clerk Maxwell, professor at Cambridge University, predicted from theory the possibility of wireless waves some years before they were discovered. It was, in fact, Maxwell's work which led Hertz to attempt to produce them.

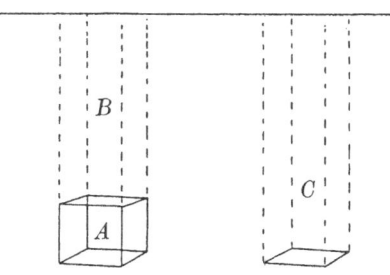

Fig. 75

Imagine a cube *A* immersed, upright, in a liquid (see **Fig. 75**).

The downward force on the top of the cube is equal to the weight of the column of liquid *B*.

The upward force on the bottom of the cube will be equal to the weight of liquid which would stand on the bottom of the cube if the rest of the cube were removed, i.e. the weight of the column of water *C*.

∴ Upthrust = weight of column of water *C*
   − weight of column of water *B*
   = weight of cube of water of same volume as *A*
   = weight of water displaced.

Q.E.D.

Calculation of density of iron and paraffin from a previous experiment.

Turn back to the readings of the weights of the iron cube in air, water and paraffin. From these readings the density of iron and paraffin can be calculated.

Upthrust of water on iron cube = 8 gm.

∴ By Archimedes' Principle,

Weight of water displaced	= 8 gm.
∴ Volume ,, ,,	= 8 c.c.
∴ Volume of iron cube	= 8 c.c.
But Weight of iron cube	= 62 gm.
∴ Density of iron	$= \frac{62}{8} = 7\cdot7$ gm. per c.c.
Again, Weight of paraffin displaced	= 7 gm.
But Volume of paraffin displaced	= volume of cube = 8 c.c.
∴ Density of paraffin	$= \frac{7}{8} = 0\cdot87$ gm. per c.c.

What becomes of the loss in weight?

One of the conundrums which is said to have interested ancient philosophers is the problem of the fish and the bucket. A fish floating in a bucket of water must weigh nothing, otherwise it would sink; that is to say, if it were attached by string to a balance the reading of the balance must be zero. If, then, the weight of a fish in water is zero, does the weight of the bucket and its contents increase when the fish is put into it?

The simplest way of solving the problem is to perform an actual experiment. Weigh a bucket containing water, drop in a fish and reweigh.

The following experiment will do equally well.

Suspend a lump of metal from a spring balance and read its weight. Place a can containing water on a compression spring balance and obtain its weight. Now immerse the lump of metal in the water and read the two balances again.

Here is a typical set of results.

Weight of metal in air	= 140 gm.
,, ,, water	= 120 gm.
∴ Loss in weight of metal	= 20 gm.
1st weight of can and water	= 500 gm.
2nd ,, ,, ,,	= 520 gm.
∴ Gain in weight of can and water	= 20 gm.

Thus the loss in weight of the metal is exactly equal to the gain in weight of the can and water. Therefore, when a fish is put into a bucket of water the weight of the bucket and its contents will increase by exactly the weight of the fish.

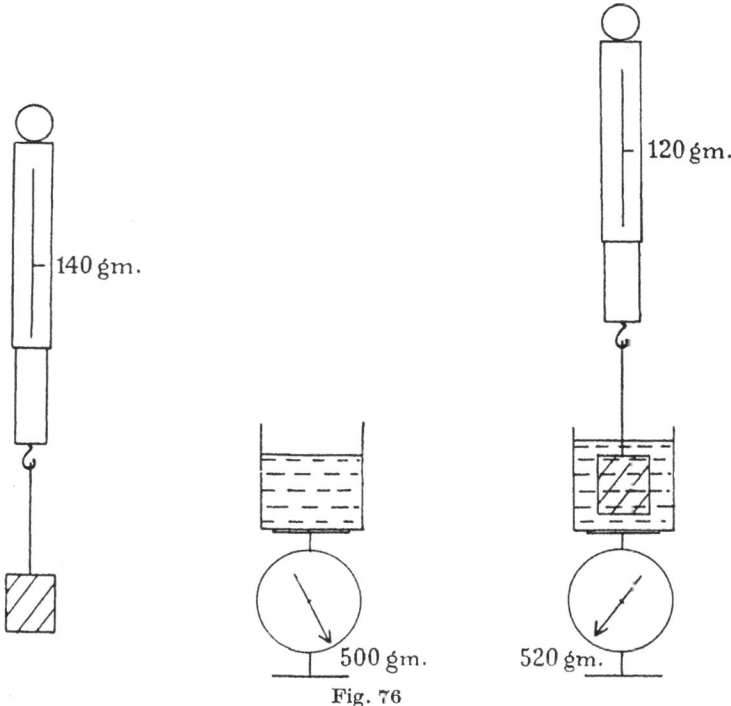

140 gm.

120 gm.

500 gm.

520 gm.

Fig. 76

Speaking generally, the loss in weight of a substance immersed in a liquid is caused by the liquid supporting part of the weight, and this results in an equal increase in weight of the liquid and vessel containing it. We shall see later that this is an example of the law, "action and reaction are equal and opposite" (see p. 117).

Floating bodies.

Suppose a block of wood, suspended by string from a spring balance, is gradually lowered into water. As the wood becomes more deeply immersed in the water the reading of the spring balance will decrease, for the weight of water displaced and consequently the upthrust of the water on the wood is increasing. Eventually there will come a time when the reading of the spring balance is zero. The wood is now floating and its entire weight is being supported by the upthrust of the water.

Thus when a body floats it displaces a weight of liquid exactly equal to its own weight. Iron, even when completely immersed, displaces a weight of water less than its own weight, and consequently sinks.

Suppose a block of wood floats in water with exactly half its volume submerged. The density of the wood must be half the density of water, 0·5 gm. per c.c. For the weight of water displaced by the wood is equal to the weight of the wood, but has only half its volume.

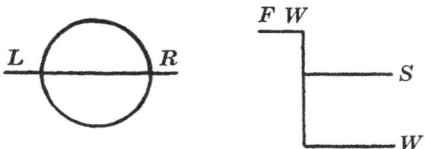

Fig. 77. Plimsoll line.

Ships which are made largely of iron float because they are hollow, and can displace a weight of water equal to their own weight. The White Star liner *Majestic* weighs 56,500 tons, and must therefore displace 56,500 tons of water.

When a ship is loaded with passengers or cargo its weight increases, and it sinks deeper into the water to displace an extra weight of water equal to the extra load.

Years ago ships were lost at sea because they were overloaded, and in 1890 Samuel Plimsoll, a native of Bristol, prevailed upon Parliament to make a law that all ships should be marked with a safety line, now known as the Plimsoll line. To-day all ships bear this line, so that it is easy to see when their water displacement is approaching its safety limit.

There are usually several lines on the side of the ship, the chief ones being shown in Fig. 77. LR stands for Lloyd's Register, the mark of the great shipping insurance firm. FW is the fresh-water line, and S and W the lines for Summer and Winter on the sea respectively. Storms are less severe on fresh-water lakes than on the open sea, and at sea they are less severe in Summer than in Winter. Hence the distance from the waterline to the deck, known as the freeboard, can be increasingly great in these three cases.

The submarine.

One of the most interesting examples of the application of Archimedes' Principle is the submarine.

The submarine is a boat, in shape rather like a fish, which can be sealed up and made to dive under the surface of the water.

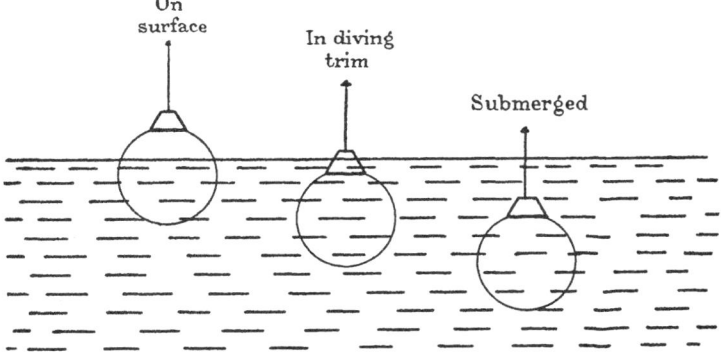

Fig. 78. Submarine.

In order to submerge she has large tanks at the bow and stern, and also smaller auxiliary tanks on either side amidships. These can be filled with water, and thus increase her weight. When they are practically full she is almost completely submerged, and is said to be in "diving trim".

Besides an ordinary vertical rudder she has horizontal rudders called hydroplanes. By inclining these hydroplanes downwards while she is moving and in diving trim, she can submerge completely and maintain her depth. The hydroplanes enable her to steer up and down just as the vertical rudder enables her to steer sideways.

When submerged, the submarine's periscope, which is a special kind of telescope, pokes up above the surface, and facilitates navigation.

During the Great War, when the German submarines were sinking Allied ships faster than they could be replaced, a device was perfected, called the hydrophone, by means of which a distant submarine could be detected by the noise of her propeller.

As soon as she knew that a destroyer was on her trail, a submarine would let more water into her tanks to make her heavier than the water she displaced, and sink to the bottom, where she could switch off her engines. There she would rest, hoping anxiously that the destroyer would not find her position, and that the deadly under-water bombs, or depth charges, would be dropped too far away to do her any damage.

In deep water, however, a submarine dare not sink to the bottom, as the pressure would tend to crack her like an egg.

If she escaped, she could rise once more to the surface when the danger had passed, by blowing the water out of her tanks with compressed air.

A submarine is propelled on the surface by Diesel oil engines, but since they use up much air while working, electric motors are used for under-water propulsion. The electricity is supplied from storage batteries which have to be charged at intervals by the Diesel engines.

Her chief weapon is the torpedo, which is itself like a tiny submarine filled with explosive. It can be launched through a tube while the vessel is submerged; it has its own propellers and rudders, and a neat little engine driven by compressed air. When its nose strikes anything, a piece of steel called the pistol is pushed on to the detonating charge, and it explodes with great violence.

In the early days of submarines 200 tons was thought to be a large displacement. The modern British submarine X_1 has a displacement of 3600 tons, carries a crew of 120 men, and can cruise on the surface at a speed of 20 knots.

In many ways, however, size is a disadvantage. Large submarines are more difficult to handle than smaller vessels. They have to take in so great a weight of water to submerge that there is always the danger of them getting out of hand, and plunging violently to the bottom. Moreover, smaller vessels have a better chance of coming unseen upon an enemy ship.

Floating docks.

A floating dock is a flat ship which can lift another ship up on its back. There are floating docks so large that they can raise even the greatest liners out of the water.

Topical Press Agency

Fig. 79. The *Majestic* in floating dock.

They consist essentially of a vast tank divided into compartments which can be filled with water and emptied again by compressed air. Above the surface are two walls which support the lifted liner, and prevent it toppling over (see Fig. 79).

The compartments are first filled with sea water so that the dock sinks deeply and the ship to be raised can sail, or be drawn by tugs, between the walls of the dock. The water is then blown out of the compartments gradually, and the dock rises until it begins to support the ship. Eventually when the ship is high and dry, the dock must displace a weight of water equal to its own weight plus that of the ship.

At Southampton may be seen the largest floating dock in the world. It is 960 ft. long, and has a lifting power of 60,000 tons. This dock can lift the *Majestic* clean out of the water in 5 hours— "the world's greatest weight-lifting feat".

The Singapore dock belonging to the British Admiralty, which was towed from the Tyne to Singapore in 1928, is nearly as big. It is 855 ft. long and lifts 50,000 tons.

Balloons and airships.

An airship floats in the air just as a submarine floats in the sea. We have already mentioned that Archimedes' Principle applies to gases as well as to liquids.

In order to float in air, a balloon must be filled with a gas lighter than air, so that the upthrust of the air may be sufficient to support both the weight of the gas and the rest of the balloon. The first balloon was made by the Montgolfier brothers. They knew that hot air rises, and so made their balloon of paper open at the bottom, and beneath this hole they lighted a fire.

In 1783, an actual flight of 25 minutes' duration was made by two Frenchmen, in a hot air balloon beneath which a fire in a brazier was suspended. Two years later, Blanchard made his famous crossing of the Channel in a hydrogen balloon, rowed through the air by two large oars. Most modern airships are filled with hydrogen, which is the lightest gas known, being 14·4 times lighter than air.

Airships began to be provided with engines when the internal combustion engine had been perfected. They were called dirigibles because their motion could be directed. (French *Diriger*, to direct.) Also, in place of the non-rigid gas bag, they were given a light framework of aluminium or steel, inside which a number of separate gas bags or ballonets were placed. Airship development, of which the greatest pioneer was Count von Zeppelin, is recent history. In 1919, the British airship R 34 made the first return flight across the Atlantic; and in 1926

Captain Amundsen flew across the North Pole in the "Norge". In 1929 the Graf Zeppelin flew round the world in 21 days, and now makes regular trips between Germany and South America.

Most readers of this book will be familiar with the tragic story of the R 101, the greatest airship ever built. She left Cardington,

From a model in the Science Museum, South Kensington
Fig. 80. The first hydrogen balloon of Charles.

Bedfordshire, for India, in the evening of October 4th, 1930, with nearly all Britain's experts in airship design and navigation on board.

It was a dark night as she crossed the Channel, and the heavy rain on her huge bulk must have considerably increased her weight. She was observed to be flying very low over France, and at Beauvais, at about 1 a.m., her nose dipped suddenly; she hit a hill-side, and burst into flames. Scarcely any of her passengers or crew escaped.

A commission appointed to enquire into the causes of the disaster found that an increase in her size had made her unstable, and her gas bags were leaking at a greater rate than was known.

She was 730 ft. long, and had a capacity of 5,000,000 cu. ft. She carried five engines of 600 H.P. and her accommodation for

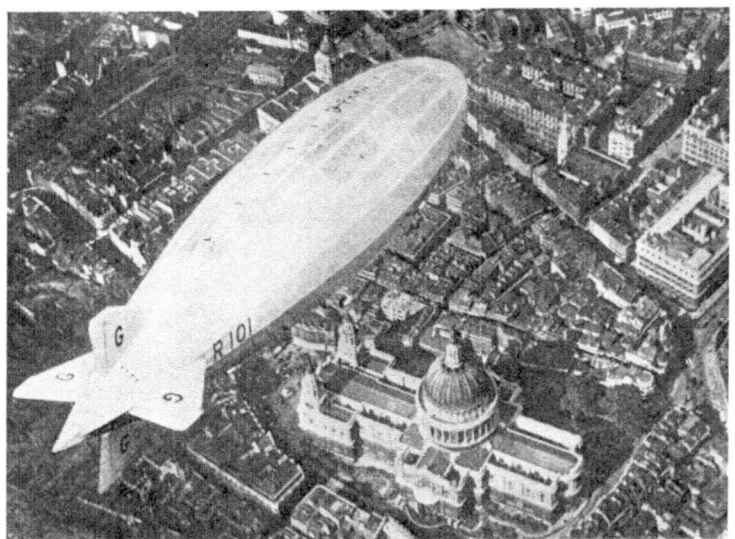

Photo: *Aerofilms, Ltd.*

Fig. 81. The R 101 flying over London.

100 passengers was fitted with the luxury one associates with a great liner.

While this book was being written, another airship disaster, comparable in magnitude to that of the R 101, occurred. The United States Navy dirigible "Akron" plunged into the sea during a storm, taking down with her all of her crew but three.

To find the lifting power of the R 101.

Volume of airship   = 5,000,000 cu. ft.
Density of air   = 0·08 lb. per cu. ft.

By Archimedes' Principle,

Upthrust of air on airship	= weight of air displaced.
∴ Upthrust	$= 5,000,000 \times 0.08$ lb.
	$= 400,000$ lb.
Density of hydrogen	$= 0.0056$ lb. per cu. ft.
∴ Weight of hydrogen filling airship	$= 5,000,000 \times 0.0056$
	$= 18,000$ lb.
∴ Lifting power of the airship	$= 400,000 - 18,000$
	$= 382,000$ lb.
	$= 170$ tons.

This lifting power has to support the weight of the fabric, the framework, engines, etc., in fact, the weight of the whole ship apart from the hydrogen. It will be realised that not many tons could be spared for passengers and their effects, despite ruthless efforts to keep everything as light as possible.

Another gas, with which airships can be filled, helium, has the great advantage of being non-inflammable. It is twice as heavy as hydrogen, but we shall see that its lifting power is not very much less. Unfortunately it is very rare. If the R 101 had been filled with helium instead of hydrogen, there would probably have been no loss of life when she was wrecked.

Lifting power of R 101 *if filled with helium.*

Volume of airship	$= 5,000,000$ cu. ft.

As before

Upthrust of air	$= 400,000$ lb.
Density of helium	$= 2 \times$ density of hydrogen.
∴ Weight of helium filling ship	$= 2 \times$ weight of hydrogen
	filling ship
	$= 2 \times 18,000$ lb.
	$= 36,000$ lb.
∴ Lifting power of airship	$= 400,000 - 36,000$
	$= 364,000$ lb.
	$= 162.5$ tons.

The lifting power when the ship is filled with helium is only 7·5 tons less than when she is filled with hydrogen. The sacrifice of this amount of buoyancy, although by no means negligible, is a price worth paying for safety from fire, and there is no doubt that all airships would be filled with helium if only sufficient were available.

Limit to which a balloon can ascend. As a balloon or airship rises the pressure of the air gets less. The buoyancy therefore decreases, since the weight of air displaced is less. The hydrogen, also, in the balloon tends to expand, and for this reason the gas bags are not usually filled to capacity at ground level. They are equipped with valves to enable the hydrogen to escape when its pressure becomes dangerously greater than that of the atmosphere. In order to rise as high as possible an airship must throw out all her water ballast. The maximum height to which a large airship such as the R 34 can rise is about 14,000 ft.

In 1932 Monsieur Piccard made an ascent in a balloon to a height of $10\frac{1}{2}$ miles. The pressure of the air at this height is about one-eighth of its value at sea level. In order to preserve his buoyancy, Monsieur Piccard made his balloon so that it could expand enormously. Its volume had to increase eight times at the height of $10\frac{1}{2}$ miles in order that the upthrust of the air might be the same as at sea level.

Sounding balloons, carrying self-reading instruments to register temperature and pressure, have been sent up to heights as great as 20 miles. They are made of rubber, and burst when the pressure of the hydrogen becomes considerably greater than the pressure of the atmosphere. The height reached can be calculated from the recorded value of the atmospheric pressure. They drop quite gently like a parachute, and the instruments are enclosed in a bamboo framework to protect them when they reach the ground. A label is attached informing the finder that he is entitled to a small reward on taking the instruments to a Post-office.

Experiment 1. *To find the density of a solid using Archimedes' Principle.* Weigh accurately the solid in air; weigh it in water suspending it from one arm of the balance in a can of water supported on a bridge across the pan of the balance.

Work out your result by a method similar to that used on p. 100.

Experiment 2. *To find the density of a liquid using Archimedes' Principle.* Weigh some convenient solid, such as a glass stopper, in air, water and the liquid. The apparatus required will be the same as that in Experiment 1.

Fig. 82. Piccard's balloon. At a height of ten miles its volume increased eight times and it became spherical owing to the decreased atmospheric pressure.

Work out the density of the liquid by a method similar to that used on p. 100.

Experiment 3. *To find the density of a solid which floats, using Archimedes' Principle.* This is a good exercise in accurate weighing and clear thinking.

Since the solid floats, it has to be immersed by tying to it a sinker—a piece of iron or lead, for example. Only three weighings need to be made, the weights of the solid in air, of the sinker in water, and of the sinker and solid in water.

Try and work out the density of the solid from these readings.

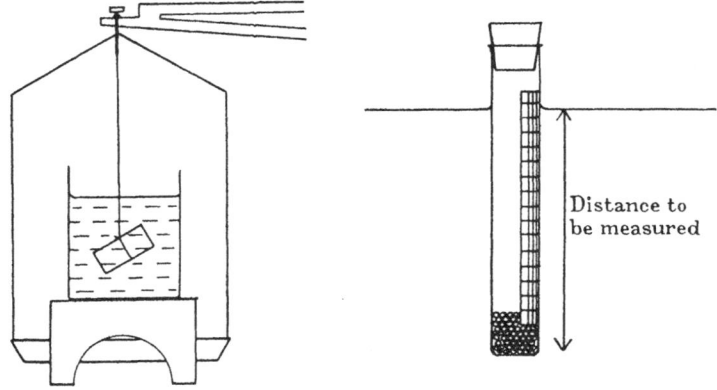

Fig. 83   Fig. 84. A home-made hydrometer.

Experiment 4. *To make a hydrometer from a test tube and hence find the specific gravity of a liquid.* Heat the closed end of a test tube and when it is soft press it on a piece of asbestos so that the end is flattened. The test tube now has an approximately uniform cross-section.

Put enough lead shot into the test tube so that it floats about three-fourths immersed in water. Fasten some graph paper inside to act as a scale and insert a cork.

Float the test tube in water and then in the liquid, and note the positions of the surface levels on the scale (see Fig. 84). Then

$$\text{Specific gravity of liquid} = \frac{\text{Length of tube immersed in water}}{\text{Length of tube immersed in liquid}}.$$

Prove this, explaining each step in your argument.

Summary

The Principle of Archimedes.

When a body is totally or partially immersed in a fluid it experiences an upthrust equal to the weight of fluid displaced.

This principle can be verified experimentally by the cylinder and bucket experiment, and it can also be proved theoretically from the fundamental laws of pressure in liquids.

When a body floats it displaces a weight of fluid equal to its own weight.

The submarine, floating dock and airship are examples of the application of Archimedes' Principle.

The density of a solid and a liquid may be found by weighing the solid in air, water and the liquid, and applying Archimedes' Principle.

QUESTIONS

1. State Archimedes' Principle.

A cube of side 2 cm. is immersed upright in water with its top face 10 cm. below the surface of the water. Calculate the force of the water on the top and bottom faces, and hence the upthrust. Use this to prove Archimedes' Principle. Find also the force of the water on the other faces.

2. Explain the following, noting in particular the physical principles involved:

(*a*) When a swimmer is being hauled into a boat from the water he appears to get heavier as he leaves the water.

(*b*) When ice, which is floating on the surface of water in a bucket, melts, the level of water in the bucket does not rise.

(*c*) The weight of a body in a vacuum is different from its weight in air. (Will it be greater or less? Why?)

(*d*) A bird is placed in a sealed box. Will the weight of the box be greater when the bird is standing on the bottom or flying in the air?

(*e*) A ball of lead and a ball of wood are hung from the ends of a pivoted beam, and found to balance exactly in air. What will happen when the whole apparatus is placed in a box which is evacuated?

3. What fraction of the volume of an iceberg shows above the surface of the sea, given that the density of ice is 55 lb. per cu. ft., and that of sea water 64 lb. per cu. ft.? Explain your method fully.

4. A boy floats in water and 0·95 of his volume is submerged. If he weighs 7 stone what is his volume? (Density of water = 62½ lb. per cu. ft.)

5. A piece of iron weighing 220 gm. floats in mercury (sp. gr. 13·6) with five-ninths of its volume immersed. Find the density and the volume of the iron.

6. (a) A lump of metal weighs 54 gm. in air and 34 gm. in water. Find the density of the metal.

   (b) If it weighs 38 gm. in paraffin, what is the density of paraffin?

7. A piece of lead appears to weigh 100 gm. when suspended in water. What does it weigh in air? (Density of lead = 11·3 gm./c.c.)

8. A block of cork (density 0·20 gm. per c.c.) weighed in the ordinary way on a balance appears to weigh 10 gm. Taking the density of air as 0·0012 gm. per c.c., what is the weight, allowing for the upthrust of the air? Even this is not strictly accurate: what other similar error should be allowed for, and how will it affect the answer?

9. A canal barge 32 ft. long and 6 ft. broad is 3 ft. out of the water. It is loaded until it is only 1 ft. out of the water. What is the load in tons? (Density of water = 62½ lb. per cu. ft.)

10. A string which breaks under a tension of 7 lb. wt. is used to support an aluminium block weighing 9 lb., which is suspended in a bucket of water. If the block is raised slowly out of the water, what fraction of the volume will have emerged before the string breaks? (Sp. gr. of aluminium = 2·7.)

11. What must be the volume of the chambers of a floating dock which can raise a liner weighing 60,000 tons? (Density of sea water = 64 lb. per cu. ft.)

12. A piece of cork weighs 5 gm. in air. It is sunk in water by tying it to a piece of metal which weighs 35 gm. in water, and their combined weight is 20 gm. Find the density of the cork.

13. A life-belt has a volume of 1½ cu. ft., and its density is one-fourth that of sea water. What weight is necessary to immerse it completely in the sea? (Density of sea water = 64 lb. per cu. ft.) Do you think it is adequate to save a 15 stone man from drowning?

14. A piece of glass of density 2·5 gm. per c.c. is found to require a downward force of 15 gm. wt. to keep it below the surface of mercury (sp. gr. 13·6). Calculate the volume of the glass and its weight in air.

15. A hollow brass sphere appears to lose half its weight when weighed in water. What fraction of the sphere is hollow? (Density of brass = 8 gm. per c.c.)

16. The R 34 had a capacity of 2,000,000 cu. ft., and her weight was 59 tons (including fuel). Taking the density of air as 0·08 lb. per cu. ft., find her (extra) lifting power in tons.

17. For the sport of balloon jumping, a person requires a balloon with a lifting power 7 lb. wt. less than his own weight. His effective weight is then only 7 lb., and he can take huge leaps into the air. Find what volume of hydrogen balloon two boys *A* and *B* of weight 10 and 7 stone respectively will each require.

If *B* takes *A*'s balloon by accident, how much gas must he let out before he can descend? (Ignore the weight of the fabric.)

1 cu. ft. of hydrogen weighs 0·0056 lb. per cu. ft. Air has a density 14·4 times that of hydrogen.

18. What happens to a balloon when it ascends to great heights? What determines the maximum height to which it can ascend?

School Certificate Questions

19. A submarine, sailing in fresh water with the top of its tower on a level with the surface of the water, passes into the open sea. Describe, giving reasons for your answer, what will happen to the submarine.

If the volume of the whole submarine is 7000 cu. ft., find what change must be made in the weight of the water in the water compartments in order that it may leave the tower just on a level with the open sea surface.

Density of fresh water = 62·5 lb. per cu. ft. Sp. gr. of sea water = 1·024.

20. A beaker of water stands on a spring compression balance which reads 170 gm.; a stone suspended by a thread is now lowered into the water until just fully immersed; the reading is now 190 gm.; finally the stone is allowed to rest on the bottom of the beaker, the reading being 220 gm. Account for the variation in these readings and find the specific gravity of the stone.

21. State and explain the principle of Archimedes.

A vessel is partly filled with liquid of density 2·5 gm. per c.c. and partly with water. The liquids do not mix. A solid object floats with 70 per cent. of its volume immersed in the water and the remainder in the other liquid. Find the density of the solid.

22. A cube of wood of volume 27 c.c. and sp. gr. 0·75 is placed gently in a beaker filled with a liquid of sp. gr. 0·95 so as to float on it. What volume of liquid will overflow and what depth of wood will be submerged?

23. What weight of wire (of sp. gr. 8) must be wrapped around a cork (of sp. gr. 0·25 and of weight 7 gm.) in order just to make it sink in water?

24. A block of wax which is known to contain a piece of lead weighs 1102 gm. when suspended in air, but only 102 gm. when suspended in water. Find the mass of the embedded piece of lead. (Sp. gr. of lead and wax, 11 and 0·9 respectively.)

25. Describe the common hydrometer and explain the method of graduating it.

The stem of a hydrometer is graduated upwards from 0 to 100 in equal parts. The volume of the graduated stem is one-quarter of the volume of the instrument below the 0 mark. When placed in water it sinks to the 20 mark. Find the density of the liquid in which it sinks to the 0 mark.

26. A common hydrometer floats in water with 2 cm. of its stem above the surface. When floating in a liquid of sp. gr. 1·2 it has 22 cm. of its stem above the surface. How much of the stem will be above the surface if it is placed in a liquid of sp. gr. 1·1?

27. A cylindrical tube, closed at the lower end, floats upright in oil of sp. gr. 0·90. The length of the tube immersed is increased by 5 cm. when a piece of metal is placed in the tube, and by 4·5 cm. when the piece of metal is attached to the bottom of the tube on the outside. Find the sp. gr. of the metal. [1st M.B.]

SECTION II. MECHANICS

Chapter VI

THE LEVER. CENTRE OF GRAVITY

Force.

Mechanics is the study of forces. It is a branch of science which is closely related to everyday experience, because on everyone and everything, even when at rest, forces are acting.

If the chair in which you are sitting is drawn suddenly from under you, you fall to the floor. This is because the earth is attracting you downwards with a force which is called your *weight*.

While the chair is supporting you, it is exerting an upward force on you which is exactly equal and opposite to your weight. Your downward weight (the force of the earth on you) and the upward force of the chair on you are balancing one another; they are said to be *in equilibrium*.

If the chair is rickety and incapable of exerting an upward force equal to your weight, it collapses and you descend. On the other hand, a dentist occasionally finds it convenient for his chair to exert an upward force on you greater than your weight, and you rise. In neither of these two cases are the forces in equilibrium.

Now if two boys sit on one chair it exerts an upward force equal to their combined weight. But it never exerts this double force on one boy: if it did there would be some competition as to who got up first, since the remaining boy would be thrown into the air. The chair adjusts itself to the person's weight as long as it is capable of so doing, but it never shows its capacity for exerting superior force by thrusting anyone into the air. Similarly, if you press on a wall it does not knock you over. It presses backwards with an equal and opposite force as long as it is strong enough to stop you going through. Newton was the first man to observe this law, and put it into words. He said, "*Action and reaction are equal and opposite*".

The measurement of force.

In order to measure forces it is necessary to select a unit. For many purposes it is convenient to compare all forces with the force of the earth's pull upon the standard lb. or with the force of the earth's pull upon the standard gm.; **forces are measured,** therefore, either **in lb. wt. or gm. wt.**

A *spring* (i.e. a coiled wire) has a characteristic which enables forces to be measured conveniently. When it is extended by exerting a force on one end, the extension is proportional to the force. If, for example, a force of 3 lb. wt. extends the spring 1 in., 6 lb. wt. will extend it 2 in., and so on.

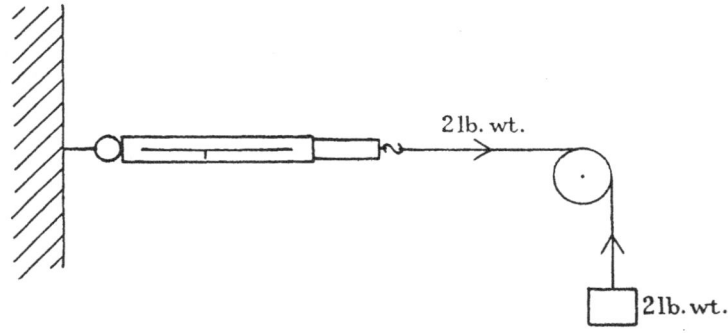

Fig. 85

The spring balance consists of a spring having one end fixed, and a hook on the other end. There is a scale by the side of the spring to show the extension, but instead of being marked in inches or centimetres, it is marked in lb. wt. or gm. wt., having been tested by the manufacturers to find the relation between the force and the extension. It is very simple to make a spring balance of one's own in the laboratory, and calibrate it so that it reads directly in lb. wt. or gm. wt.

The spring balance is generally used for measuring the force of the earth's pull upon the object which hangs from its hook, i.e. for measuring weight. Alternatively, it can be used for measuring a man's strength, in which case he pulls upon the hook, and compares his force with the force of the earth's pull upon various weights.

Tension in a string. For simple experiments in the laboratory, we often allow the earth to exert additional forces on bodies by hanging weights on them. If it is desired to apply a force in a direction other than vertical, then the weight is hung at the end of a string attached to the body, and passing over a well-pivoted pulley. We can show by a very simple experiment that the pull or tension of the string is equal to the weight hanging from it. Attach a spring balance to a hook in the wall (see Fig 85), and tie to it a string which passes over a pulley and carries a 2 lb. wt. Then wherever the pulley is placed to alter the inclination of the string, the reading of the spring balance will be 2 lb.

One end of the string pulls the spring balance with a force of 2 lb. wt. The other end of the string must be pulling upwards with a force of 2 lb. wt., otherwise the weight would drop to the ground. The ends of a string which is in tension always pull with equal and opposite forces.

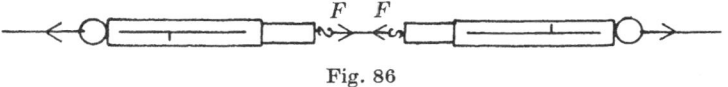

Fig. 86

As an experimental proof of this fact, tie a string to two spring balances (see Fig. 86) and pull outwards with both hands. The force at each end of the string is balanced by an equal and opposite force in the spring balance. But the readings of the two balances will be found to be identical. Therefore the two forces in the string are equal. The string is in equilibrium.

Machines.

One of the problems to which man has been turning his wits from earliest times is how to increase the comparatively small force which he can exert with his muscles into forces large enough to give him mastery over his environment. An apparatus for increasing a force (or for applying a force more conveniently) is known as a *machine*.

We who are growing up in this age of machinery are apt to take as matter-of-fact its staggering achievements—a crane which can pick up 350 tons, a gun which can throw a shell 60 miles, or a seaplane that can move through the air at 400 m.p.h.

The striking feature of the history of machines is that it is nearly all compressed into the last hundred years. Nevertheless, simple machines which are the essential components of all others —the lever, the wheel, the pulley, the inclined plane and the screw—have been known and used for centuries. From experiments with them, the fundamental laws of mechanics have been discovered, and the knowledge has been used in the development of more elaborate machines.

The lever.

The lever is one of the simplest machines, and an important member of many more complicated machines. It consists of a bar free to move over a pivot.

A crowbar is a simple lever. It is used to dislodge heavy paving stones, the edge of a neighbouring stone being used as a pivot. The burglar's jemmy is a crowbar, collapsible so that it can be concealed about his person. He pushes one end under the door or window he wishes to prise, and pivots his jemmy on the surrounding woodwork.

A cyclist's tool bag usually contains three tyre-lifting levers. There are other levers in his machine itself, the cranks, which connect the pedals to the wheel over which passes the chain, enabling the force of the foot on the pedal to turn the wheel, and the brake handles.

The turning effect of forces. The value of a lever depends upon its length. Next time you open a sardine tin, slip one end of a strong metal bar through the key, and see how much easier it is to prise off the lid by turning the end of the bar than by twisting the key with the thumb and finger. A burglar's jemmy is collapsible because a short crowbar would be of very little use to him. Similarly, a door would be very much harder to open and shut if the knob were situated only an inch or two from the hinge.

Thus the turning effect of a force depends not only on the size of the force itself, but also on its distance from the pivot. Archimedes made a remark which expresses vividly the fact that given a lever long enough, an enormous weight can be lifted by a very small force. He said, "Give me a place to stand on, and I will move the world".

Now the see-saw is a form of lever. To see-saw successfully, two boys must be almost exactly balanced. Therefore the heavier boy gets closer to the pivot, while the lighter one moves further away. Then they find that the turning power of one is practically the same as the turning power of the other, and it is a muscular thrust of the feet on the ground which causes one to go up, and the other down.

This suggests a simple experiment for the investigation of the turning effect of forces.

Hang two unequal weights on a bar pivoted at its centre (see Fig. 87), and measure their distances from the pivot (called the *fulcrum*) F, when their turning effects are equal, i.e. when the

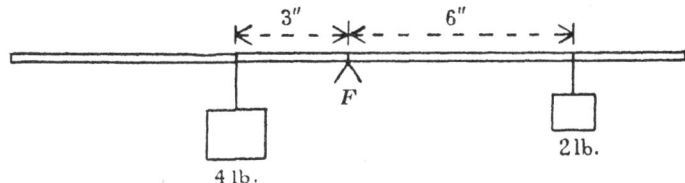

Fig. 87

bar remains horizontal. The following are readings obtained in this way:

Distance of 2 lb. wt. from $F = 6$ in.
Distance of 4 lb. wt. from $F = 3$ in.

The hanging weights are merely convenient methods of applying known forces. It will be seen that one force is double the other, and its distance from the fulcrum is exactly half.

Now we know that the turning powers of the forces are equal, since the bar remains horizontal; but the forces multiplied by their distances from the fulcrum are also equal.

Right-hand force × distance = $2 \times 6 = 12$ lb. in.
Left-hand   force × distance = $4 \times 3 = 12$ lb. in.

This suggests that the turning power of a force may be measured by the product of the force and its distance from the fulcrum. A special name has been given to the product. It is known as the *Moment* of the force.

Thus,   *Moment of a force = Force × distance from fulcrum.*

Whatever weights and distances are taken in the above experiment, the moments of the two forces are equal.

Hence the *Law of the Lever* may be expressed

Moment of Right-hand force = Moment of Left-hand force

when the forces are balanced, i.e. in equilibrium.

General principle of moments.

If several forces acting on a bar are in equilibrium, it is found that the sum of the moments tending to turn the bar in one direction (e.g. the direction in which the hands of a clock rotate—

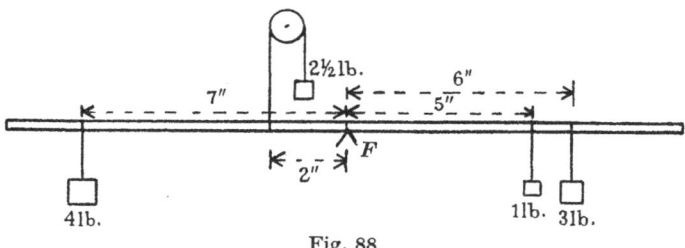

Fig. 88

clockwise) is exactly equal to the sum of the moments tending to turn it in the opposite direction (anti-clockwise). Taking the readings from Fig. 88:

Clockwise moments $= (1 \times 5) + (3 \times 6) + (2\frac{1}{2} \times 2) = 28$ lb. in.
Anti-clockwise moments $= (4 \times 7)$ $= 28$ lb. in.

∴   **Clockwise moments = Anti-clockwise moments.**

Moment of an oblique force.

It sometimes happens that a force is not applied at right angles to a lever. In this case the moment of the force is not so great. The cyclist's foot only exerts a maximum turning effect on the crank of a bicycle when the crank is horizontal. When the crank is vertical a downward pressure of the foot exerts no turning effect at all; in this position a boy can stand on the pedals, and they will not move. In the intermediate positions the moment depends on how obliquely the foot exerts its force on the crank.

Suppose a boy exerts a steady downward force of 40 lb. wt., on one pedal of a bicycle, the length of the crank being 7 in. We will find the turning effect or moment of the force when the crank is in three positions, (a) horizontal, (b) vertical, (c) at 45° to the horizontal.

(a) Moment when the crank is horizontal $= 40 \times 7$
$$= 280 \text{ lb. in.}$$

(b) When the crank is vertical there is no leverage for a vertical force to exert any turning effect.

Moment when crank is vertical $= 40 \times 0$ lb. in.
$$= 0.$$

(c) It is tempting to guess that the moment when the crank is at 45° to the horizontal will be half the value when it is

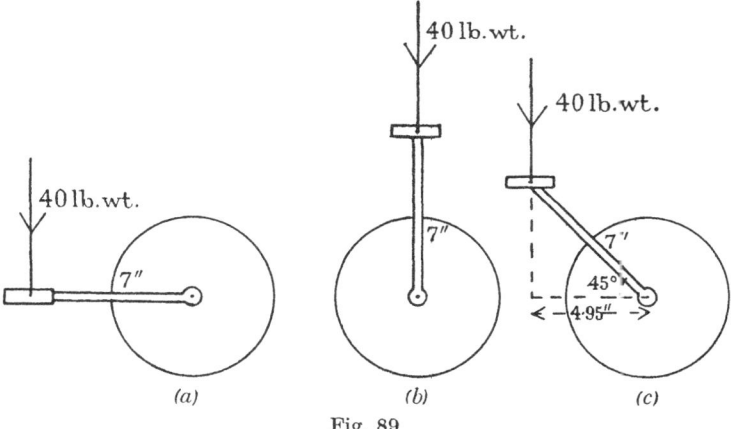

Fig. 89

horizontal, i.e. $\frac{280}{2} = 140$. The guess, however, is incorrect. The effective turning arm of a vertical force is its horizontal distance from the pivot. Thus we must draw a figure accurately to scale and measure the horizontal distance from the axle to the 40 lb. wt. This distance is 4·95 in. (see Fig. 89).

$\therefore$ Moment when crank is at 45° to horizontal
$$= 40 \times 4 \cdot 95 = 198 \text{ lb. in.}$$

The horizontal distance taken as the effective arm in this case is the perpendicular distance from the pivot to the force.

Speaking generally, **the moment of an oblique force is the product of the force and the perpendicular distance from the fulcrum to the line of action of the force.**

An experiment to verify the principle of moments, employing an oblique force. Set up an apparatus similar to Fig. 90. Pivot the bar about its centre, and apply an oblique force to one end by hanging a 3 lb. wt. from a string which passes over a pulley. (The pull in the string is numerically equal to the 3 lb. wt.) Balance the bar by adjusting the position of a second force of

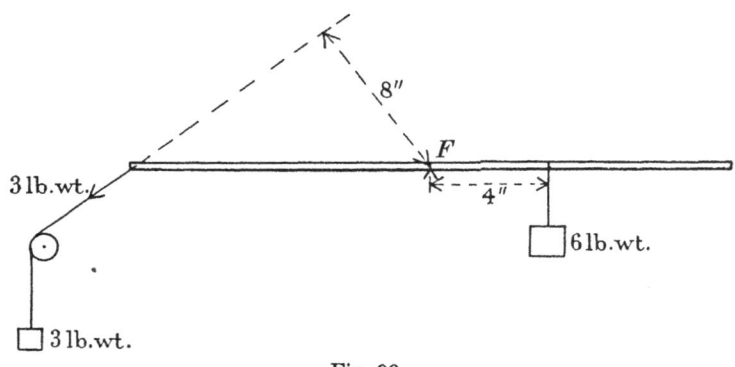

Fig. 90

6 lb. wt. Measure the perpendicular distances from the fulcrum to the lines of action of the forces (see Fig. 90).

Then   Clockwise moments $= 6 \times 4 = 24$ lb. in.
         Anti-clockwise moments $= 3 \times 8 = 24$ lb. in.

Since the principle of moments holds for this case, it is clear that our definition of the moment of an oblique force is sound.

Levers.

The chief point of interest about a simple machine is the number of times it increases the force applied—a quantity known as its **Mechanical Advantage.**

The force applied is called the **effort, *P*,** and the weight to be lifted is called the **load, *W*.** Thus if a load of 10 lb. wt. is raised by an effort of 1 lb. wt., the mechanical advantage is 10.

Definition. **Mechanical advantage** $= \dfrac{\text{Load}}{\text{Effort}} = \dfrac{W}{P}$.

Levers may usefully be divided into two types:

(*a*) those in which the fulcrum lies between the load and the effort,

(*b*) those in which it lies at one end of the lever.

Levers of the first type. The mechanical advantage of the lever in Fig. 91 clearly depends on the lengths of *BF* and *AF*. Suppose

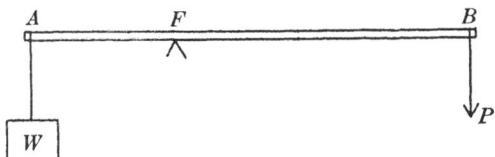

Fig. 91

the effort P is just sufficient to balance the load W. Then a minute increase in P will cause the load to be raised.

By the law of the lever,

Moment of W about $F =$ Moment of P about F,
$$W \times AF = P \times FB,$$
$$\frac{W}{P} = \frac{FB}{AF}.$$

$\therefore$ Mechanical advantage of lever $= \dfrac{W}{P} = \dfrac{FB}{AF}.$

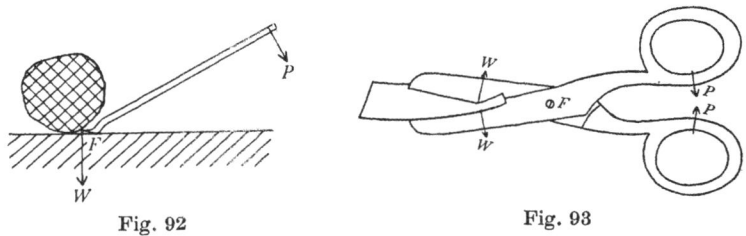

Fig. 92   Fig. 93

Examples of this type of lever are the crowbar, the hand-brake of a motor car, and a pair of scissors (see Figs. 92 and

93). The beam of the common balance is also a lever of this type.

There is another type of balance, known as the steelyard, whose beam is not pivoted at its mid-point (see Fig. 94). A simple steelyard is often used by butchers for weighing carcases too heavy for the ordinary scales. The body to be weighed, *W*, is hung from the hook and the beam is made horizontal by moving along the sliding jockey weight *P* which, having a greater lever-

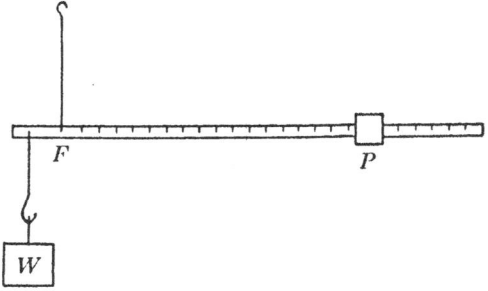

Fig. 94

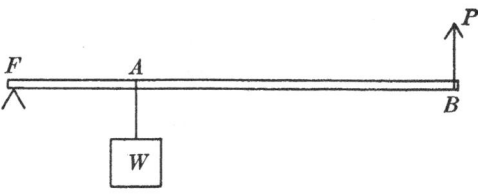

Fig. 95

age, is considerably lighter than *W*. This is also the type of balance used by doctors and schools for weighing people. The fulcrum *F* is usually supported on a vertical pillar and the hook is connected by a vertical rod to a platform on which stands the person to be weighed. From the end of the long arm hangs a scale pan on which may be placed small weights reading stones (because of their large leverage), while the position of the jockey *P* reads lbs.

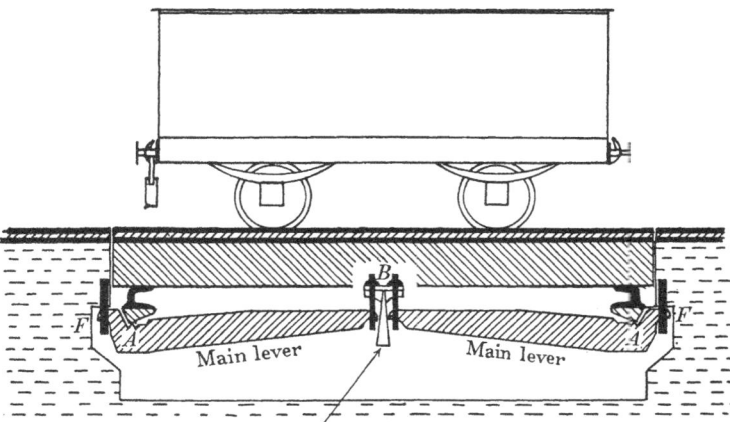

Transverse section of transfer lever supporting
the end of the two main levers

Fig. 96 (a). Weighbridge—side section.

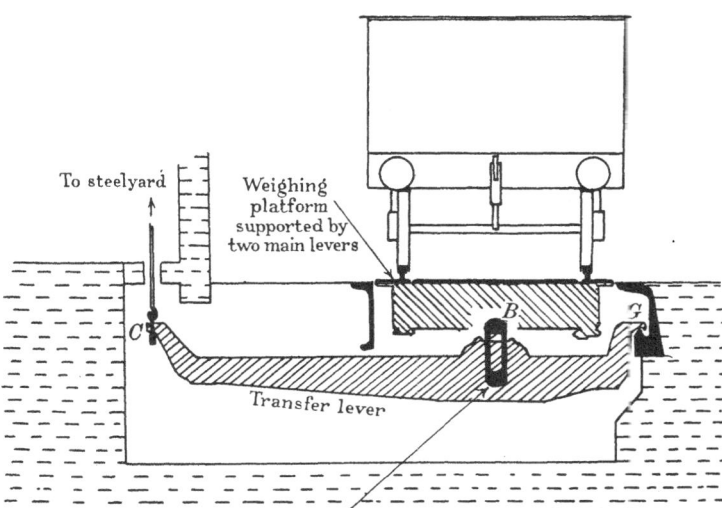

Link supporting end of main lever

Fig. 96 (b). Weighbridge—end section.

Levers of the second type. Fig. 95 shows a lever of the second type. When the lever is balanced,

Moment of W about F = Moment of P about F,

$\therefore$ $W \times AF$ $\qquad\qquad = P \times BF.$

$\therefore$ Mechanical advantage $= \dfrac{W}{P} = \dfrac{BF}{AF}.$

It is possible to use a crowbar as a lever of the second type if the ground is used as the fulcrum. The main and transfer levers of a weighbridge (see Fig. 96 (a) and (b)) are levers of this type. Weighbridges are commonly used by railway companies for weighing loaded trucks and by coal merchants for weighing loads of coal. The truck is run on to a platform, when its weight bears down on the main levers at the points marked A in Fig. 96 (a). The main levers (between them) communicate a much smaller force than the load at B to the transfer lever, since B is considerably further from the fulcrum F, than A. The transfer lever is shown in Fig. 96 (b). Its fulcrum is at G, the main levers apply a load to it at B, and it is supported by a rod at C, which is connected to a steelyard. Both the transfer lever and the steelyard have a considerable mechanical advantage with the result that a load weighing many tons (in some cases 100 tons) may be balanced by comparatively small weights on the steelyard.

Another example of a lever of the second type is the oar of a boat. One might imagine the oar to be a lever of the first type, with the fulcrum at the rowlock. However, the rowlock moves as a part of the boat, and it is on the rowlock that the oar exerts the force which overcomes the resistance of the boat, and drives it forward. The resistance of the boat corresponds to the load W. The fulcrum is at the blade, supposed at rest in the water (although this is not strictly true), and the effort is applied at the handle.

Again, the wheelbarrow works on the lever principle, for when the handles are lifted, a force much less than the weight of the barrow and its contents is required. The fulcrum is the axle of the wheel. A pair of nutcrackers is a double lever of this type.

Fig. 97 shows a lever of the second type which has a mechanical disadvantage.

If the lever is balanced,

Moment of W about F = Moment of P about F,

∴ $W \times AF \qquad = P \times BF.$

∴ Mechanical advantage $= \dfrac{W}{P} = \dfrac{BF}{AF}.$

Since BF is less than AF, the mechanical advantage is less than 1. Thus the effort required must be greater than the load.

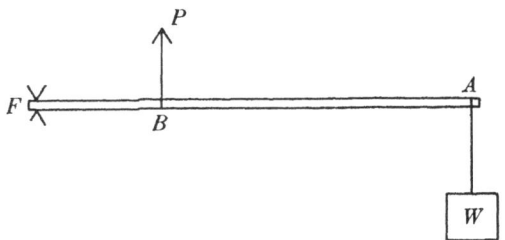

Fig. 97

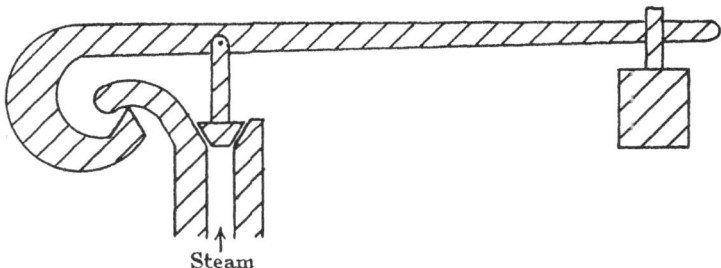

Fig. 98. The safety valve of a steam engine.

This is sometimes distinctly useful. For example, the boiler of a steam engine must be supplied with a safety valve to enable the steam to blow off when its pressure becomes too high. A crude safety valve might take the form of a hole in the boiler over which is placed a large weight. When the pressure of the steam reached a value enabling it to exert a greater upward force than the weight over the hole, the weight would be lifted, and the steam would escape until the pressure had dropped sufficiently.

However, by using a lever similar to that shown in Fig. 98, a considerably smaller weight, *W*, may be used.

The human forearm, when being raised, is another example of a lever of this kind (see Fig. 99). The fulcrum is at the elbow, and when a weight, held in the palm of the hand, is lifted, the biceps muscle, which is attached to the forearm at a distance equal to only one-tenth its length, exerts the pull. The strength of the pull has to be ten times the weight lifted. The advantage here, of course, is economy of space. If the biceps muscle were attached near to the wrist, the arm would swell enormously when the hand is raised. When the arm is being straightened out, the forearm acts as a lever of the first order with a mechanical advantage considerably less than

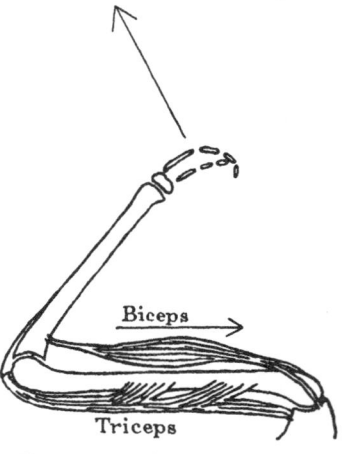

Fig. 99. The human forearm.

1. The triceps muscle exerts the pull as shown in Fig. 99. At all other joints in the body there are levers operated by muscles.

The wheel.

The wheel is a special form of lever. Fig. 100 shows two equal weights passing over a pulley wheel and supporting each other.

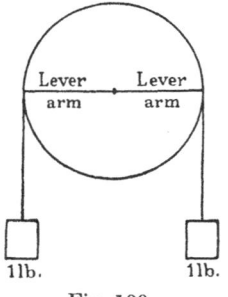

Fig. 100

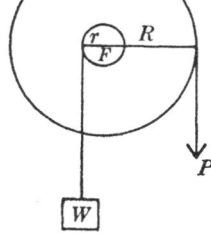

Fig. 101

The fulcrum is the axle of the wheel, and the arms cf the lever are both equal to the radius. The mechanical advantage is thus 1.

The wheel and axle.

The wheel and axle consist of a wheel mounted on an axle of smaller radius, round both of which ropes can be coiled in opposite directions (see **Fig. 101**).

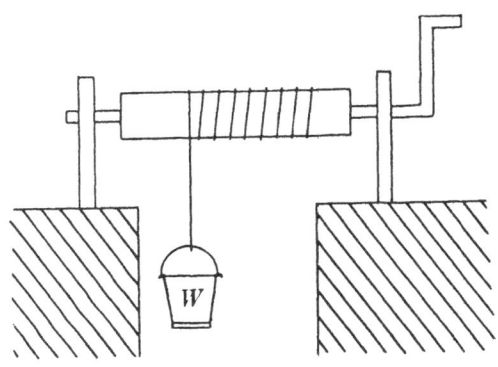

Fig. 102. Simple winch.

Let   Radius of axle $= r$,
                             Radius of wheel $= R$.

Taking moments about fulcrum F,

$$Wr = PR.$$
$$\therefore \ \frac{W}{P} = \frac{R}{r}$$

= Mechanical advantage.

A common form of the wheel and axle is the simple winch (see Fig. 102), which is used for drawing buckets of water up from a well.

The steering wheel of a car is another example—the larger the wheel, the greater the leverage for the driver. If the steering wheel came off, it would require enormous strength to steer the car by grasping the steering shaft.

Centre of gravity.

The point about which a body balances is called its *centre of gravity.* In the case of a ruler or any uniform rod, it is the mid-point.

The whole weight of a body may be considered as concentrated at and acting through its centre of gravity. Consider the case of the

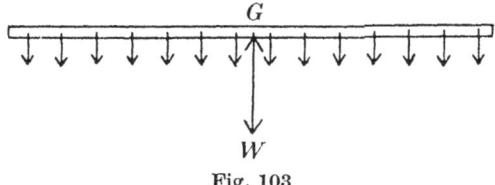

Fig. 103

uniform rod balanced at its centre on a knife edge. The weight of the rod is distributed along its length, and is made up of the sum of the weights of every tiny particle of substance in the rod. This is represented in Fig. 103, the arrows standing for the weight of the particles, but there should, of course, be millions more. The net-turning effect about G of all these tiny forces is zero, and therefore if we are to represent them all by a single force equal to their sum, W, the force must pass through G in order to have no moment about G.

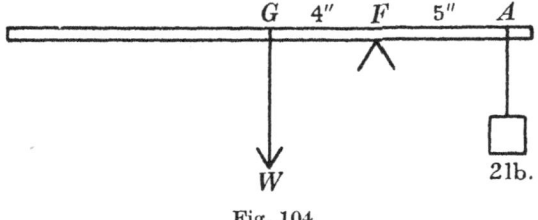

Fig. 104

If a rod is balanced on a fine steel point, the centre of gravity of the rod must lie vertically above the point, but it is inside the bar. To define centre of gravity as the point about which a body balances is therefore unsatisfactory. We define it as follows:

The centre of gravity of a body is the point at which its whole weight may be considered to act.

To find the weight of a rod without using a balance. The centre of gravity *G* of the rod is first found by balancing it on a knife edge; it will be at the centre if the rod is perfectly uniform.

It is then pivoted about some point *F* (see Fig. 104), and is made horizontal by adjusting the position *A* of the 2 lb. wt. *GF* and *AF* are measured.

Suppose
$$GF = 4 \text{ in.,}$$
$$AF = 5 \text{ in.}$$

Taking moments about *F*,
$$W \times 4 = 2 \times 5.$$
$$\therefore \quad W = \tfrac{10}{4} = 2\tfrac{1}{2} \text{ lb.}$$

Reactions of supports.

(The rest of this chapter may, with advantage, be omitted at the first reading.)

Suppose a bar of weight 10 lb. is supported horizontally on two knife edges. The two supports must push up on the bar with a total force of 10 lb. wt. (otherwise the bar would fall or rise), and if they are placed at equal distances on either side of the centre of gravity of the bar they will each exert a force of 5 lb.

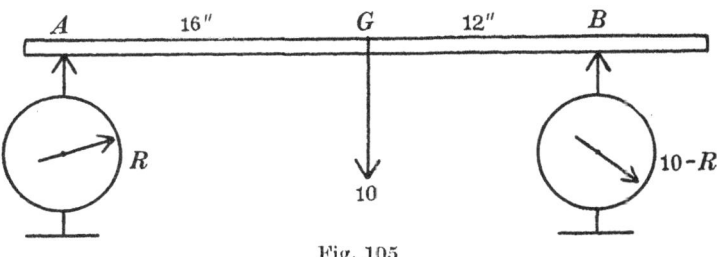

Fig. 105

wt. This can be verified by placing the supports on compression spring balances.

Suppose, however, one of the supports is placed at a distance of 16 in. from the centre of gravity *G*, and the other at a distance of 12 in. from *G*. What are the reactions now?

Let   *R* lb. wt.   = reaction of one support.
Then   (10 − *R*) lb. wt. = reaction of other support.

The moments of the reactions about G must be equal, other-wise the rod would turn about G.

Taking moments about G,

$$16R = 12 \ (10 - R).$$
$$\therefore \ 28R = 120,$$
$$R = \tfrac{120}{28} = 4\tfrac{2}{7} \text{ lb. wt.},$$
$$10 - R = 5\tfrac{3}{7} \text{ lb. wt.}$$

Again this result may be verified by experiment.

Here we have a body in equilibrium under the action of three forces. Since it has no tendency to turn about any point we could have taken moments about any point, and the clockwise and anti-clockwise moments would have been equal, i.e. the algebraic sum of the moments would have been zero.

Another convenient point about which to take moments is one of the points of support, since the moment of the force at this point will then be zero.

Taking moments about B,

$$28R = 12 \times 10,$$
$$R = \tfrac{120}{28} = 4\tfrac{2}{7} \text{ lb. wt.}$$

The result is the same as before.

Again, if a known weight were hung on the bar we could cal-culate the reactions at the supports by including its moment in our equation. In this way an engineer can calculate the forces on the supports of a bridge when a heavy lorry is passing over it.

An experiment to find the centre of gravity of a triangle.

Cut out a large cardboard triangle, preferably with unequal sides, so that it is not a special type.

Hang it up vertically by a pin stuck through near to one corner, and allow it to hang perfectly freely by making the hole a little wider than the pin. Hang also a plumb-line of fine thread from the pin, and mark its position on the cardboard with a pencil at two points as far apart as is convenient. Take great care over this; use a set square if it is a help. The centre of gravity of the triangle will lie vertically below the point of support (why?) and hence somewhere on the plumb-line.

Repeat the experiment with the triangle hanging from points

near the other two corners. Take down the triangle and draw on it with a fine pencil the three directions of the plumb-line. The centre of gravity must lie on all three lines and hence they should intersect in one point. How nearly they do this will be an indication of the accuracy with which the experiment has been performed.

Now turn the triangle over, and on the other side join the corners to the mid-points of the opposite sides. These three lines, called the medians, should also meet in a point. Stick a pin through this point, and see how near it is to the centre of gravity. Try to explain what you find.

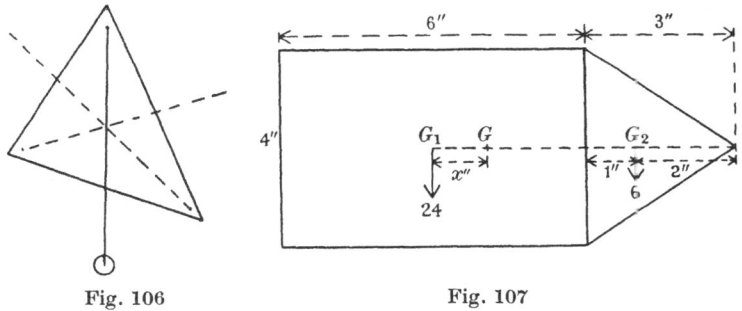

Fig. 106   Fig. 107

Calculation of the position of the centre of gravity.

A thin plate or card of uniform thickness is known as a *lamina*. From the experiment just described it can be shown that the centre of gravity of a triangular lamina is at the point of intersection of the medians, i.e. one-third of the way up each of them.

Example 1. To find the centre of gravity of the lamina shown in Fig. 107. The centre of gravity of the rectangle is at its centre G, and that of the triangle at G_2, one-third way up the median.

Suppose 1 sq. in. of the lamina has unit weight.

The weight of the rectangle will be 24 units, and that of the triangle $= \frac{1}{2} \times 4 \times 3 = 6$ units.

Let G be the centre of gravity of the whole lamina. G_1G_2 may be regarded as a weightless lever, with weights of 24 and 6 units at its ends, which balances about a pivot at G, the point through which the resultant weight acts.

Taking moments about G,
$$24 \times G_1G = 6 \times G_2G,$$
i.e.
$$24x = 6 \ (4-x),$$
$$30x = 24,$$
$$x = \tfrac{4}{5} \text{ in.}$$

Example 2. To find the centre of gravity of a piece of uniform wire, ABC, bent in the form of an **L**. The centres of gravity of AB and BC are at their mid-points, G_1 and G_2, respectively. G_1G_2

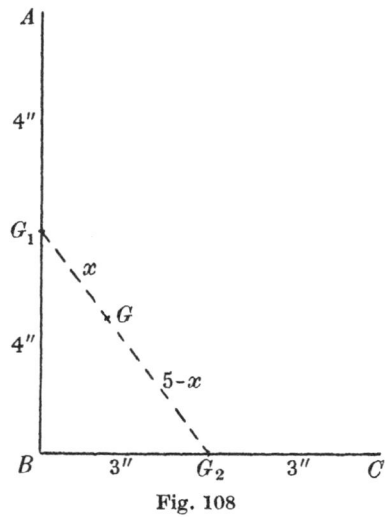

Fig. 108

may be regarded as a weightless lever with weights of 8 and 6 units acting at its ends. G, the point about which this lever would balance, is the centre of gravity of ABC.

Taking moments about G,
$$8 \times G_1G = 6 \times G_2G,$$
i.e.
$$8x = 6 \ (5-x), \quad G_1G_2 = \sqrt{4^2+3^2}$$
$$= 5 \text{ in.}$$
$$\therefore \quad x = \tfrac{30}{14}$$
$$= 2\tfrac{1}{7} \text{ in.}$$

Resultant of parallel forces.

(a) *Two like parallel forces.* Fig. 109 shows three parallel forces, P, Q and W, in equilibrium. The case is similar to that of Fig. 105.

Now P and Q could be replaced by a single upward force of $(P+Q)$, known as their *resultant*, which is equivalent to them combined. This force must act at the same point on the bar as

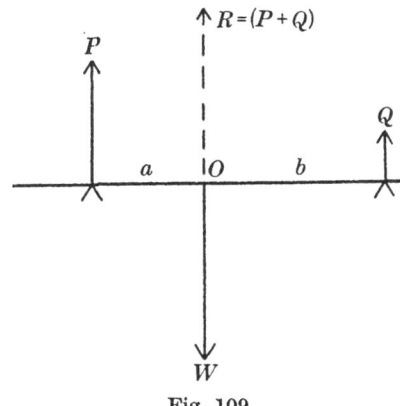

Fig. 109

W, since it must exactly balance W, and therefore act along the same line.

Thus to find the positions of the resultant of P and Q take moments about O, $Pa = Qb.$

(b) *Two unequal unlike parallel forces.* In order to balance two unequal unlike parallel forces P and Q (see Fig. 110), a downward force would have to be applied at O. OAB may be regarded as a weightless bar pivoted at O. If the pivot lay between P and Q the bar would spin. O must be outside AB and nearer to the larger force P.

Thus the resultant of P and Q $(P-Q)$, must act at O.

Taking moments about O,

$$Pa = Qb.$$

Two equal unlike parallel forces—a couple.

When two equal unlike parallel forces, known as a *couple* or *torque* (Fig. 111), act on a bar it is impossible to balance the bar by

MECHANICS

applying a single force. If a pivot were placed between the forces the bar would spin, and if it were placed beyond one of the forces the moments of the two forces about the pivot could never be equal.

Thus a couple cannot be represented by a single force: it has no resultant. It can be balanced only by another couple.

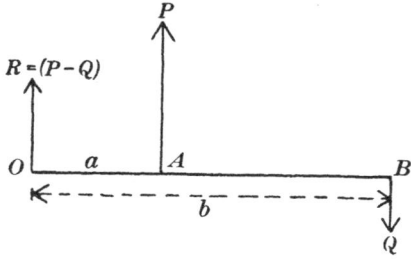

Fig. 110

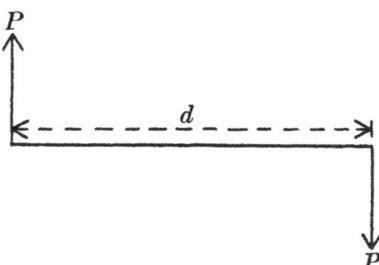

Fig. 111. A couple.

The effect of a couple is rotary only. It has no capacity for causing or changing motion in a straight line.

Thus a compass needle, on which the earth's field exerts a couple, is turned until it points N. and S., but has no tendency to move bodily either N. or S.

The moment of a couple is the same about any point, namely, one of the forces × the perpendicular distance between them. (Take a point along the bar produced, distant x from one of the forces, and show that the moment about this point is Pd.)

Stability. The tendency of a body, such as a bus, to topple

when tilted depends on the position of its centre of gravity.
A top-heavy vehicle is unsafe because there is a tendency to tilt

By courtesy of the London General Omnibus Co.

Fig. 112. Tilting an omnibus to test the position of its centre of gravity.
The upstairs seats alone are loaded with sandbags (in place of passengers)
so that the bus shall be as top-heavy as it is ever likely to be in practice.

when taking corners. Buses and charabancs are therefore designed to have as low a centre of gravity as possible, and the
London General Omnibus Company test the position of the

centre of gravity of their buses by tilting them as shown in
Fig. 112. An indicator is attached to the bus to record the angle
of tilt, and the bus is prevented from crashing right over by
ropes.

You can perform a similar simple experiment to investigate
the condition for toppling. Take a rectangular block of wood and
mark the height of its centre of gravity G in the middle of one of
its faces. Draw a chalk mark from G to one of the corners A.
Stand the block on a drawing board, putting A against a ruler

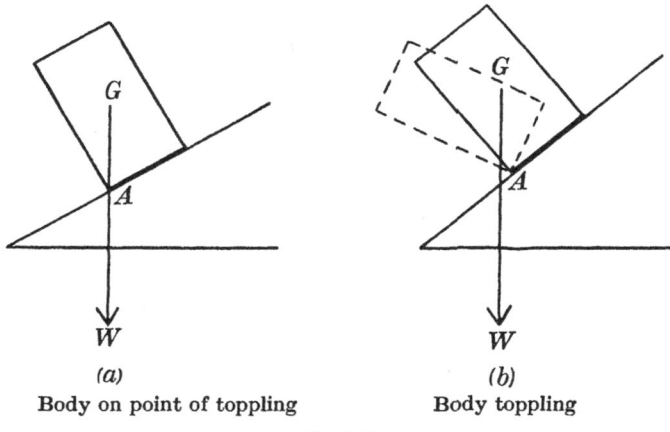

| (a) | (b) |
| Body on point of toppling | Body toppling |

Fig. 113

so that the block will not slide, and gradually tilt the board until
the block topples about A. You will find that the block is just on
the point of toppling when GA is vertical (see Fig. 113 (a)). If
the vertical through the centre of gravity falls outside A (see
Fig. 113 (b)) the block will topple because its weight has a turning
effect about A in an anti-clockwise direction. If the vertical through
G cuts the base of the block inside A the turning effect of the
weight about A tends to restore the block if it is rotated slightly
about A.

Consider Fig. 113 (a). Suppose the centre of gravity G were
lowered. It is clear that the vertical through the new centre of
gravity would fall well within A and consequently the block

could be tilted still further before toppling. Thus a body with a low centre of gravity is more stable than one with a high centre of gravity.

It is dangerous in a rough sea to stand up in a rowing boat, for by so doing the centre of gravity of the boat and its occupants is raised. It is then far less stable, and liable to be overturned by a wave.

The most interesting example of precarious equilibrium is the tight-rope walker. To prevent himself from toppling he must keep his centre of gravity vertically over the rope, and to help him to do this he often carries a long pole or a weighted umbrella. When he feels himself falling to one side he pushes the pole or umbrella over slightly to the other side, and thus restores the position of the common centre of gravity. The celebrated Blondin used to cross the Niagara falls on a rope, and perform some theatrical feat, such as cooking and eating an omelette, half-way across. To take a less exciting but equally instructive example, a boy walking on a wall holds his arms extended, and draws in one if he feels himself falling to that side, thus bringing his centre of gravity once more above the wall. A man carrying a heavy portmanteau in one hand, holds the other arm outwards for a similar reason, and the coal-man carrying a sack of coal on his back leans forward to prevent himself from toppling over backwards.

Types of equilibrium.

There are three types of equilibrium, *stable, unstable* and *neutral.* If a body, when slightly displaced, tends to return to its

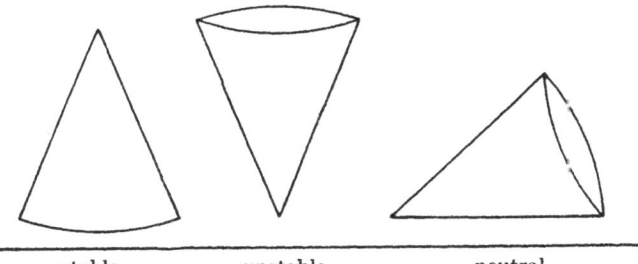

stable   unstable   neutral
Fig. 114. Types of equilibrium.

original position, it is said to be in stable equilibrium. If it tends
to upset completely, it is in unstable equilibrium, and if it has
no tendency either to return or become displaced further, it is in
neutral equilibrium. Simple examples of these three types of
equilibrium are a cone standing on its base, its tip, and its side,
respectively (see Fig. 114).

The fundamental difference between these types of equilibrium
lies in the behaviour of the centre of gravity. A body always
tends to have its centre of gravity at the lowest possible point.
When a body in stable equilibrium is displaced, its centre of
gravity rises, and tends to fall again, thus restoring the body to
its original position. When a
body in unstable equilibrium is
displaced, its centre of gravity
falls, and tends to continue to
fall, thus totally upsetting the
body. On the other hand, on
moving a body in neutral equilibrium such as a ball, the centre
of gravity neither rises nor falls.

A toy called a Kelly is a good
example of stable equilibrium.
It has a rounded base which is
heavily weighted, and in consequence, however much knocked
about, always returns to an upright position. Another interesting toy is a bird with a weighted

Fig. 115

tail which cannot (except by a violent blow) be knocked off its
perch. Cut one of these birds from cardboard, and weight its
tail with drawing pins (see Fig. 115). Try and find the position of
its centre of gravity. You may make the interesting discovery
that its centre of gravity lies outside the cardboard.

The balance.

Fig. 116 shows a balance pivoted at F. If the pans are of equal
weight why does the beam always come to rest in the horizontal
position instead of being in equilibrium in any position, like an
ordinary bar?

The answer is that the centre of gravity of the balance is not at F but at G and the balance sets itself in such a position that G is vertically below F. The equilibrium is stable, since when the beam is displaced, G rises. If G were at F, the equilibrium would be neutral, and if G were above F, unstable.

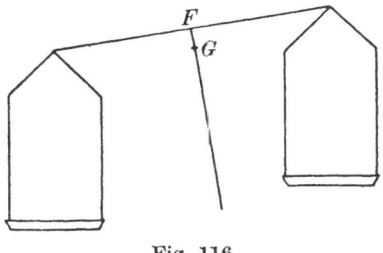

Fig. 116

The chief requisites of a balance are that it should be sensitive and rapidly attain equilibrium. For great sensitiveness FG must be short so that the moment of the weight of the balance itself shall be as small as possible. On the other hand, for rapidity of weighing FG should be large. A compromise, governed by the use to which the balance is to be put, is therefore necessary.

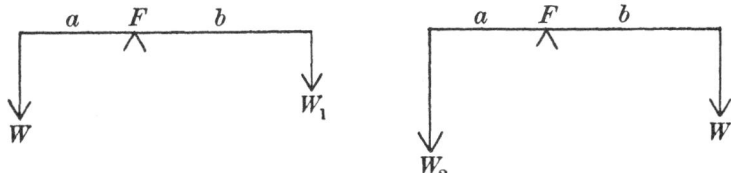

Fig. 117. Double weighing.

Inaccurate balances. Double weighing. A balance may have arms of unequal length and yet swing true owing to the arms being unequally loaded. The correct weight of an object may be found by weighing it alternately in each pan.

Suppose the lengths of the arms are a, b, the correct weight of an object W, and the weights required to balance the object when placed in each pan, W_1 and W_2.

Taking moments about F,

$$Wa = W_1 b,$$
$$W_2 a = Wb.$$
$$\therefore \frac{W}{W_2} = \frac{W_1}{W}.$$
$$\therefore W = \sqrt{W_1 . W_2}.$$

Summary

The turning effect of a force is called its **Moment**. It is measured by **the product of the force and the perpendicular distance from the fulcrum to the line of action of the force.**

When a body is in equilibrium under the action of a number of forces, the sum of the clockwise moments about the fulcrum or any point is equal to the sum of the anticlockwise moments, i.e. **the algebraic sum of the moments is zero.**

The mechanical advantage of a machine $= \dfrac{\text{Load}}{\text{Effort}}.$

Levers are simple machines, and may be divided into two types, according to the position of the fulcrum.

The wheel is a special form of lever. Its mechanical advantage is 1, but can be increased by using it in the form of a wheel and axle.

The centre of gravity of a body is the point at which its whole weight may be considered to act.

There are three types of equilibrium, stable, unstable and neutral. If bodies exhibiting these types of equilibrium are slightly displaced, the centre of gravity of the first body rises, that of the second falls, and that of the third remains at the same height respectively.

A body tends to topple when the vertical through its centre of gravity falls outside its base.

QUESTIONS

1. (a) One spring balance hangs from another, and to the lower is attached a 3 lb. wt. What will be the reading of each spring balance?

(b) Two vertical balances side by side together support a 3 lb. wt. What do they read? [Check by experiment.]

2. In Fig. 118 the weight of the pulley is $\frac{1}{2}$ lb. Find the tensions in the strings A and B. State, and mark in a diagram, the forces acting on (a) the 4 lb. body, (b) the pulley.

3. Explain the following, with diagrams, marking in the fulcrum, load and effort:

(a) When using a pair of scissors it is best to insert what is to be cut near the joint.

(b) A well-designed screw-driver has a wide handle.

(c) A slogger in cricket uses "the long handle".

4. Explain, with the aid of diagrams, how the following work, and to what types of levers they belong. Mark in the fulcrum, load and effort:

(a) A hammer used as a nail extractor.

(b) A spanner.

(c) The front-wheel brake of a bicycle.

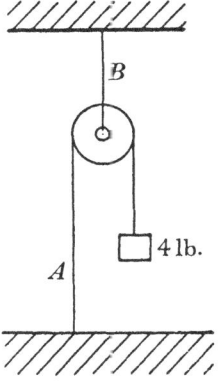

Fig. 118

5. In a steelyard for weighing meat the hook is at a horizontal distance of 2 in. from the fulcrum, and the sliding jockey weight weighs 5 lb. How far from the fulcrum must the jockey be placed to balance a weight of 30 lb. of meat?

6. A force of 60 lb. is required to crack a nut. What force must be applied to the handles of nutcrackers 5 in. long, if the nut is placed 1 in. from the hinge?

What is the mechanical advantage obtained?

7. An oar is 8 ft. long, and the rowlock is 6 ft. from the blade. Find what force it exerts on the rowlock to propel the boat when the oarsman exerts a force of 40 lb. at the end of the handle.

What is the mechanical advantage obtained?

8. What effort is required to raise a load of 40 lb. wt. by means of a wheel and axle, if the radius of the axle is 1 in. and that of the wheel 1 ft.?

9. Two boys, weighing 6 and 7 stone, sit on one side of a see-saw at distances from the pivot of 3 ft. and 5 ft., respectively. Where must a third boy, of weight 10 stone, sit in order to balance them?

10. A tramp carries his personal effects in a bundle weighing 8 lb., hanging from the end of a stick resting on his shoulder. If one-third of the stick projects behind him, what force must he exert at the other end of it to balance his load?

Neglecting the weight of the stick, what force does it exert on his shoulder?

What force would have been exerted on his shoulder if the stick had been balanced about its middle point?

11. The arms of a balance are of unequal length, the left-hand one being 16 cm. long, and the right-hand one, 15·9 cm. What will be the apparent weight of a body whose true weight is 40 gm. when weighed (*a*) in the left-hand pan, (*b*) in the right-hand pan? Work out the weights correct to 3 decimal places.

Does the average of these values give the correct weight exactly?

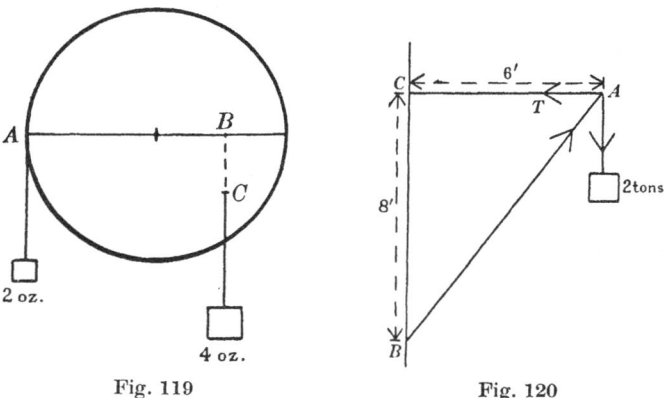

Fig. 119   Fig. 120

12. A cardboard disc (see Fig. 119) is pivoted at its centre and a 2 oz. weight hung from A. It is balanced by a 4 oz. weight hung from a point B on the same diameter. What will happen if the 4 oz. weight is attached to C, vertically below B? Give your reasons.

13. The cranks of a bicycle are $6\frac{1}{2}$ in. long. Find the moment of a downward force of 50 lb. exerted on one pedal when the crank is (*a*) horizontal, (*b*) vertical, (*c*) at 60° to the horizontal. Do part (*c*) by drawing a diagram to scale or by trigonometry and explain your method.

14. The jib AB of a simple crane is pivoted at B and a weight of 2 tons hangs from A. Find the horizontal force T to hold the jib in place, given $BC = 8$ ft., $AC = 6$ ft. (Find the moments about B of the two forces.) See Fig. 120.

15. Explain the following:

(a) It is easier to knock a boy over when he is standing on one foot than when he is standing on two.

(b) When a cyclist feels himself falling to one side, he turns his front wheel to the same side.

(c) A man who is putting the weight holds his left hand and leg extended in front of him.

(d) A racing yacht has a deep fin keel with lead ballast at its bottom.

16. A knife blade is placed under the 40 cm. mark of a uniform metre ruler and it is found that a weight of 40 gm. hung from the 10 cm. mark is required to keep it horizontal. What is the weight of the ruler?

17. A bridge 120 ft. long weighs 50 tons, and is supported at its ends. What are the reactions of the supports when a lorry of weight 10 tons is standing 20 ft. from one end?

18. A gate, 10 ft. wide and 5 ft. high, is supported by two hinges attached to one side: the weight is supported by the upper hinge and the lower one exerts a horizontal force only (this is true in practice if the hinge pins are loose in their sockets). If the gate is uniform and weighs 100 lb., show all the forces acting and determine the reactions at the hinges.

19. A wall, 7 ft. high and 15 in. thick, is built of materials weighing 120 lb. per cu. ft. A strong horizontal wind blows on it and exerts a perpendicular pressure of 45 lb. per sq. ft. Show in a diagram the resultant forces due to the wind and the weight of the wall; assuming that if the wall falls it will do so by toppling over bodily, decide whether it will or not.

School Certificate Questions

20. A platform weighing 80 lb. is held in mid-air by a man standing on the platform and holding one end of a rope which passes over a fixed pulley and has its other end fixed to the platform. If the man weighs 180 lb., with what force does he pull on the rope?

21. Explain: A ladder leaning against a smooth wall is more likely to slip when a man on it is near the top than when he is near the bottom.

22. What is the principle of moments?

A uniform beam of length 14 ft. and mass 112 lb. is used as a see-saw. If a 12 stone man sits at one end and a 4 stone boy at the other, where must the support be placed in order that the see-saw may balance?

23. Explain how it is possible to find the resultant of two unlike parallel forces which do not act at a point. Discuss the case when the forces are equal.

A uniform beam, 24 ft. long and weighing 200 lb., is suspended by two parallel ropes of equal length, one fastened 6 ft. from one end and the other 9 ft. from the other end of the beam. A man weighing 180 lb. stands at the mid-point of the beam. Calculate the tension in each rope.

24. What is meant by the moment of a force about a point?

A uniform plank AB of length 14 ft. and mass 250 lb. is supported horizontally on two trestles C and D placed 2 ft. from A and B, respectively. Find what force must be applied at B, (a) to lift the plank just clear of the trestle at C, (b) to lift the plank just clear of the trestle at D.

25. What do you understand by a couple and the moment of a couple? Devise an experiment to show that two couples of equal and opposite moments in the same plane balance each other. Draw a diagram of the apparatus you would set up, and explain the measurements you would make and the way you would use them.

26. A painter stands on a horizontal platform suspended by its ends from vertical ropes A and B 16 ft. apart. The platform weighs 50 lb. The tension in A is 140 lb. wt., and in B, 60 lb. wt. What is the weight of the painter and where is he standing?

27. A laboratory balance is known to be inaccurate owing to the unequal weights of the scale pans. Describe any way you know whereby the weight of a body could be accurately determined by it.

28. Define centre of gravity. Give examples of a body in (a) stable, (b) unstable, (c) neutral equilibrium, and explain in each case how the position of the centre of gravity determines the nature of the equilibrium.

A right circular cylinder of height 8 in. and base diameter 4 in. is placed on an adjustable inclined plane whose surface is so rough as to prevent slipping. The inclination of the plane to the horizontal is gradually increased. Find the least angle of inclination at which the cylinder topples over.

29. What is meant by the centre of gravity of a body?

A quadrilateral $ABCD$ is cut out of a uniform sheet of cardboard, the angles at A and B being right angles and the angle at C acute.

How would you find the centre of gravity of the figure (a) by geometrical construction, (b) by experiment?

30. A rectangle of sides 3 in. and 2 in. has an equilateral triangle described on one of the shorter sides. Find the centre of gravity of the whole figure so obtained.

31. The diagonals of a square of side 8 in. are drawn and one of the triangles so formed is cut out. Find the position of the centre of gravity of the remainder.

32. A cube of homogeneous material has a spherical cavity in its interior. The centre of the cavity is distant from the geometrical centre of the cube by an amount equal to 0·2 of a side and its diameter is equal to 0·5 of a side. How far is the centre of gravity of this cube from its geometrical centre?

33. A post of cross-section 6 in. square is fixed vertically and centrally to a horizontal board of the same material 2 ft. square and 4 in. thick. If the top of the post is 8 ft. above the level of the board, find the position of the centre of gravity of the structure. Find graphically the greatest angle through which the board can be turned about one of its edges so that when released it may return to its initial position.

34. Fig. 121 represents a square piece of wood, of uniform thickness, from which one-quarter has been removed. The part shown weighs

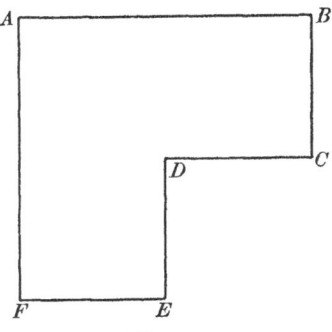

Fig. 121

100 gm. and stands on the side EF as base. Find the greatest weight which may be hung from the projecting corner at C without causing the wood to overbalance.

Chapter VII

THEORY OF MACHINES

Mechanical advantage and velocity ratio.

In the last chapter we discussed one of the simplest machines, the lever, and saw how it could magnify a force, and so give a mechanical advantage.

A small boy weighing 6 stone, and sitting on one arm of a see-saw, can lift a boy twice his weight, provided the heavier boy sits only half as far from the fulcrum. This is a lever with a mechanical advantage of 2. But the distance through which the small boy moves is twice as great as the distance through which the heavier boy moves, and as they reach the top and bottom of their swing simultaneously the small boy must on an average move twice as fast as the other. Thus the ratio of the velocities of the small boy (the effort) and the bigger boy (the load) is also 2.

In the simple weight-lifting machines used in the laboratory it is much easier to measure the corresponding distances moved by the load and effort than their velocities. Hence we take as our working definition,

$$\text{Velocity ratio} = \frac{\textbf{Distance moved by effort}}{\textbf{Distance moved by load}}.$$

In the case of a lever with a mechanical advantage of 10 the effort has to move ten times as far as the load. This is the price which must always be paid for a mechanical advantage. Any machine which is designed to have a high mechanical advantage must also have a high velocity ratio. In fact, "what is gained in force is lost in speed".

The gear box of a motor car.

An example of the operation of this rule is the gear box of a car, which enables the velocity ratio to be altered at will. Here

$$\text{Velocity ratio} = \frac{\text{Velocity of engine (crankshaft)}}{\text{Velocity of car}},$$

since the effort is applied by the engine, and the load corresponds to the resistance experienced by the car.

The idea of changing down from top to some lower gear is not merely to make the car go more slowly (most cars will go at any speed in top from 6 or 8 m.p.h. upwards), but by increasing the velocity ratio to produce a corresponding increase in the mechanical advantage. Thus in bottom gear (fast running engine and slow car) the velocity ratio is high; the mechanical advantage is therefore high also, and the

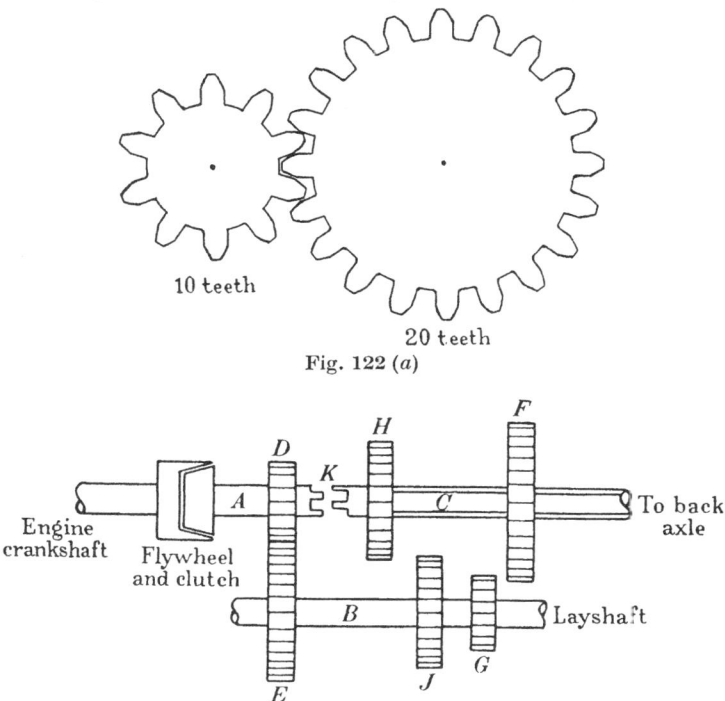

10 teeth

20 teeth

Fig. 122 (a)

Fig. 122 (b). Gears of a motor car (neutral position).

car can climb hills in this gear which are too steep for it in any other. On the other hand, for travelling along the level a high mechanical advantage is not required. Top gear therefore represents a low mechanical advantage (you cannot climb steep hills in this gear) and hence a low velocity ratio, i.e. a high speed of the car when the engine is running fast.

The gear of a car is changed by causing toothed wheels of different sizes inside the gear box to engage when the gear lever is moved. Consider a very simple example of gearing by toothed wheels. Suppose a wheel with 10 teeth coupled to the engine engages with a wheel with 20 teeth coupled to the back axle. When the smaller wheel goes round exactly once, it engages 10 teeth on the larger wheel, and turns it

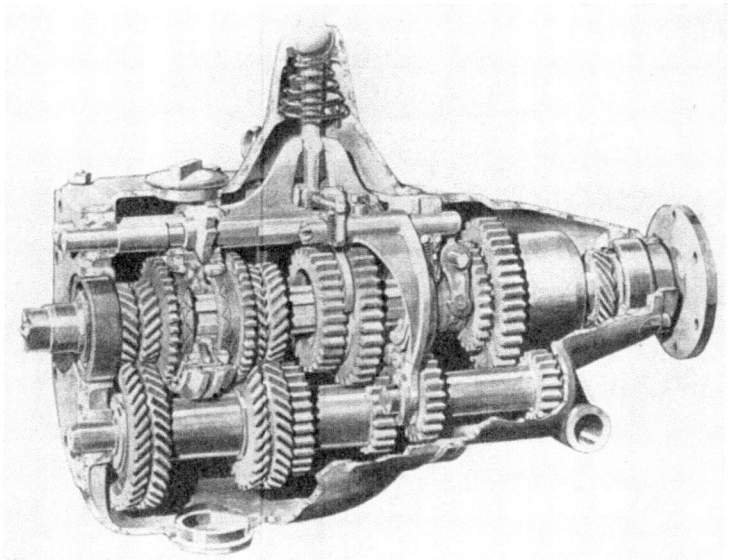

By courtesy of Morris Motors, Ltd

Fig. 123. The synchromesh (4-speed) gearbox of a Morris car. It operates on the same principle as that described in the text but contains in addition a synchromesh unit (situated mid-way between the two screw-caps in the casing), consisting of two clutches. These enable two gear wheels to be brought to the same rotational speed before engaging, and hence ensures silent gear-changing.

through half a revolution. The smaller wheel, therefore, revolves twice as fast as the larger, and the velocity ratio is 2. Any required velocity ratio may be obtained in this way by adjusting the relative number of teeth.

Fig. 122 (*b*) is a simplified diagram of the gears of a car. The shaft *A* is connected to the engine via the clutch, a device for enabling the engine to be disengaged from *A* if desired. The cone type of clutch consists of two cones, a hollow cone in the rear face of the flywheel, and a solid cone which fits exactly into the hollow one, as a finger fits into a thimble. The flywheel rotates on the crank shaft of the engine and when the two cones are pressed together, the shaft A rotates with it. When the inner cone is withdrawn (by pressing down the clutch pedal), A ceases to rotate. The cogwheels D and E are always in mesh; D has half the number of teeth of E, and hence the lay shaft B revolves at half the speed of the engine. The shaft C, which is connected direct to the back axle, is independent of A and the cogwheels H and F can be made to slide along it by moving the gear lever (but they cannot rotate without also rotating C).

When the gear lever is in neutral, shafts A and B are rotating, but they are not coupled to C, and hence the car remains at rest. To get into bottom gear the clutch is put out in order to disconnect A and B from the engine. The wheel F is moved by the gear lever, so that it engages with G. Since F is considerably larger than G, it revolves much more slowly, and thus the back wheels go round more slowly than the engine when the clutch is let in.

For second gear, H and J are connected, F and G being first disconnected. They have approximately the same number of teeth, and since the lay shaft revolves at half the speed of the engine, the shaft C must revolve at this speed.

For top gear, a direct drive is employed, the shafts A and C being connected by the dog clutch K. In this case the back wheels and the crank shaft of the engine revolve at the same speed.

Work.

The chief function of a machine is to do work. The term "work" when used in science has a definite meaning which must be carefully defined, just as the terms "try", "foul" and "free kick" need to be carefully defined in the laws of Rugby Football. When used in a technical sense, all these words have a different meaning from that of everyday speech.

In everyday speech we say a person is doing work when he tries to solve a mathematical problem, but in the technical sense work is done only when a weight is lifted or a force moved through a distance. On lifting a weight of 1 lb. vertically through 1 ft., 1 ft. lb. of work is done. When 10 lb. wt. is lifted vertically through 3 ft., then 30 ft. lb. of work is done.

Thus $Work = Force \times distance.$

It is important to notice that, according to this definition, no work is done unless something is moved. A man may hold a sack weighing 1 cwt. perfectly still at a constant level above the ground and completely exhaust himself, but he has done no external mechanical work. A crane could have held the sack with its engines shut off for an indefinite time. Or again, if a crane attempts to lift a weight too heavy for it and fails, however hard its engines may have tried it has done no mechanical work on the weight to be lifted. Work may have been done inside the engine, the boiler may even have burst, but as far as the weight to be lifted is concerned the position is *status quo*.

There is another important point about the definition of mechanical work. An ordinary man can quite simply push a motor car along a level road, but he cannot lift it. When he is pushing it horizontally he is not overcoming its weight, but merely friction. If all friction could be eliminated, the slightest touch would set it moving.

Hence we must make our definition of work more precise.

Work = Force × distance moved along its line of action.

The line of action of the weight of a body is vertical, since you will remember that the weight is simply the vertical attraction of the earth for the body, and therefore we need to measure the vertical distance moved. If a motor car is pushed up a hill (neglecting friction this time), the work done is equal to the weight of the car multiplied by the vertical height of the hill, and not its slant length.

The Principle of Work.

A lever with a mechanical advantage of 10 must also have a velocity ratio of 10. Let us suppose that with this lever a load of 200 lb. is lifted through 1 ft.

$$\begin{aligned}
\text{Mechanical advantage} &= 10, \\
\text{Load} &= 200 \text{ lb.,} \\
\therefore \text{ Effort} &= 20 \text{ lb.} \\
\text{Velocity ratio} &= 10, \\
\text{Distance moved by load} &= 1 \text{ ft.,} \\
\text{Distance moved by effort} &= 10 \text{ ft.,} \\
\therefore \text{ Work done } on \text{ load} &= 200 \times 1 = 200 \text{ ft. lb.} \\
\therefore \text{ Work done } by \text{ effort} &= 20 \times 10 = 200 \text{ ft. lb.}
\end{aligned}$$

[The student should observe and use the technical terms, *on* and *by*.]

Thus although a larger weight is being moved by a smaller one, there is no gain in work. The work put into the machine is equal to the work done by the machine.

For many years inventors occupied themselves with the problem of producing a machine which would give out more work than was put into it. Their aim was perpetual motion, and they dreamed of a machine which once started would produce just a little extra work to prevent it from stopping until it was worn out. The most weird and ingenious contraptions were devised, but whenever models were made, they never, without any exception, worked. (Such machines as windmills and water wheels do not come under the category of perpetual motion machines, since the work is being done in these cases by wind and water. The wind and water, in their turn, however, have derived their energy from the radiation of the sun and the rotation of the earth, and neither of these is inexhaustible. The idea of the perpetual motion machine is that it shall work because of its design, and need no fuel or motive power.)

The first great modern scientists realised that all this ingenuity was misplaced, and they took as an axiom what is now one of the best established laws of science. It is known as the **Principle of Work; No machine can give out more work than is put into it.**

In actual practice the useful work got out of a machine is always less than the work put in, since some of the work is wasted in overcoming friction and lifting parts of the machine. A machine which wastes very little of the work put into it is said to have a high **efficiency**:

$$\text{Efficiency} = \frac{\text{Work got out}}{\text{Work put in}}.$$

This is commonly expressed as a percentage by multiplying by 100.

Perpetual motion machines.

Fig. 124 shows two designs for perpetual motion machines. They incorporate two of the favourite erroneous ideas of the perpetual motion seekers, and dozens of similar but more elaborate machines were invented. The first one, Fig. 124 (*a*), was supposed to give a continuous circulation of water owing to the greater weight in the wider arm, and when built on a large enough scale, the steady gush was presumably to be used to work a water wheel or a turbine. The only difficulty about this machine is that the water refuses to circulate, the level in the narrow arm remaining the same as that in the wide arm for reasons which we discussed fully in Chapter IV.

The second machine, Fig. 124 (b), is a little less naïve, but equally futile. A wheel is built with compartments divided by curved partitions and in each compartment there is a heavy ball. Owing to the curvature of the partitions the balls on the right-hand side will run away from the centre of the wheel, and those on the left towards it. Consequently those on the right-hand side will have a greater leverage,

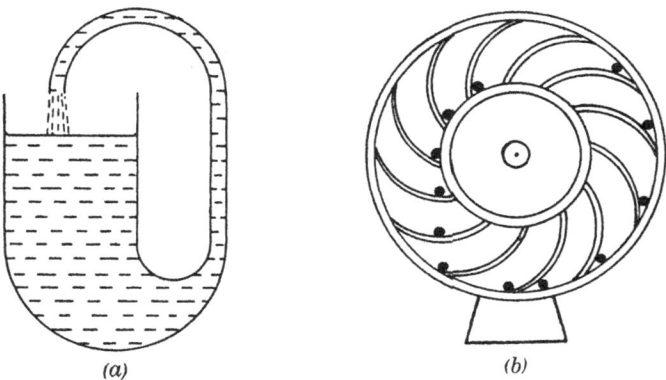

(a)   (b)

Fig. 124

and it was believed that the wheel would begin to turn in a clockwise direction. Once started, of course, it would go on for ever, for the balls on the right always run away from the centre, and those on the left towards it.

If such a machine should work, how rosy our visions of the future. A huge wheel of this design would turn day and night in every garden, and drive a dynamo to light and heat our homes, and run our domestic sewing machines and mangles all for nothing.

Unfortunately the wheel does not revolve. Although the balls on the right have a greater leverage, there are always more balls on the left, with the result that the anti-clockwise moments are exactly equal to the clockwise moments. The net turning-effect is thus nil.

Pulleys.

By means of a system of pulleys a high mechanical advantage can be obtained, and this is made use of in the hoisting gear of cranes. We shall consider first a simple pulley, and then systems or combinations of pulley wheels.

The simple pulley (see Fig. 125) consists of a wheel in a light framework, the whole being known as a pulley block. It merely serves to change the direction of a force, and has a mechanical advantage of 1. Thus a load of 4 lb. wt. requires an effort of 4 lb. wt. to balance it. The velocity ratio is also 1. A pulley is occasionally fixed to the ceiling of a warehouse to enable heavy bodies such as sacks of flour to be raised by pulling downwards on a rope passing over the wheel, instead of having to lift upwards.

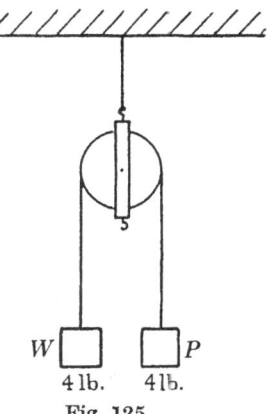

Fig. 125

If a light pulley is inverted, and a string passed round it as in Fig. 126, a 4 lb. wt. can be lifted by pulling with a force of just over 2 lb. wt., the pull being measured by a spring balance. This is because each side of the string takes half the weight, and hence half the weight is supported by the ceiling.

$$\text{Mechanical advantage} = \tfrac{4}{2} = 2.$$

However, the distance the 4 lb. wt. rises is only half the distance the effort of 2 lb. wt. moves, since when the 4 lb. wt. rises 1 ft., the string on each side must shorten by 1 ft.—a total of 2 ft. for the effort to move through. Hence the velocity ratio of the machine is 2.

For convenience, we can insert a second pulley (see Fig. 127) so that the effort is applied downwards instead of upwards. The mechanical advantage and velocity ratio are unchanged.

Calculation of efficiency of pulley system. Suppose that in the pulley system of Fig. 127 the bottom pulley was a heavy one, and that an effort of $2\tfrac{1}{2}$ lb. instead of 2 lb. was required to raise a load of 4 lb.

When load rises 1 ft., the effort must move 2 ft.

∴ Work done on load in raising it 1 ft. $= 4 \times 1 = 4$ ft. lb.

Work done by effort $= 2\frac{1}{2} \times 2 = 5$ ft. lb.

∴ Efficiency of pulley system

$$= \frac{\text{Work got out}}{\text{Work put in}}$$

$$= \tfrac{4}{5},$$

$$= 80 \text{ per cent.}$$

It may be pointed out that

$$\text{Efficiency} = \frac{W \times \text{distance moved by } W}{P \times \text{distance moved by } P} = \frac{\text{Mechanical advantage}}{\text{Velocity ratio}}$$

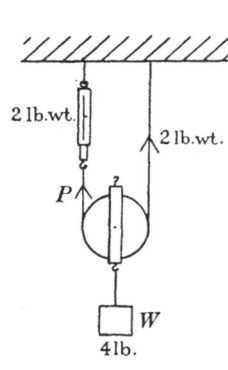

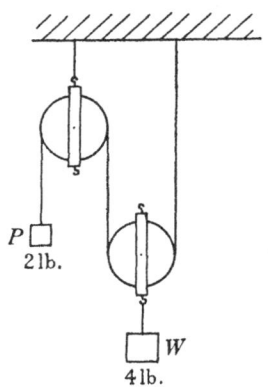

Fig. 126   Fig. 127

Thus in the above pulley system,

Mechanical advantage $= \dfrac{4}{2\frac{1}{2}} = \dfrac{8}{5}$.

Velocity ratio $\qquad = 2$.

∴ Efficiency $\qquad = \dfrac{\frac{8}{5}}{2} = \dfrac{4}{5} = 80$ per cent.

The efficiency of a simple pulley system may well be as high as this, but the efficiency of a railway engine (to take an example of a complicated machine) is much less, 15 per cent. being regarded as good. Thus 85 per cent. of the energy supplied by the coal is wasted, most of it being dissipated as heat which is lost to the atmosphere.

More elaborate pulley systems. Pulley blocks sometimes contain two or more pulley wheels side by side. Fig. 128 shows two pulley blocks, each having two wheels, connected up with one continuous string. Let us find the velocity ratio. When the load W rises 1 ft. the lower pulley block rises 1 ft., and thus there are four lengths of string which must each shorten 1 ft. Hence the effort P moves 4 ft., since it has to pull out these 4 ft. of string. The velocity ratio is therefore 4.

Let us assume the system to be frictionless, and ignore the weight of the lower pulley block (which has to be lifted). We are here dealing with a perfect machine and the principle of work becomes,

$$\text{Work done on load} = \text{Work done by effort.}$$

We will find the work done on the load when it is raised 1 ft. and by the effort moving through a corresponding distance of 4 ft.

$$W \times 1 = P \times 4.$$
$$\therefore \ \frac{W}{P} = 4.$$

Hence the *ideal mechanical advantage* of the system is 4, and is equal to the velocity ratio. In actual practice the mechanical advantage will be less than 4, since the system will not be 100 per cent. efficient.

There is another way of calculating the mechanical advantage. The load and lower pulley block are supported by four strings, which are really parts of the same string at the end of which a pull of P is being exerted. Thus the strings pull upwards with a total force of $4P$, and will be able to balance an equal downward force. If none of this upward force were needed to lift the lower pulley block or overcome friction in the pulleys, a load of $4P$ could be lifted and the mechanical advantage would be 4. In practice, only part of the upward force is available for lifting the load, and the mechanical advantage is less than 4.

The Weston pulley block.

When you next see a builder hauling up materials by hand with the aid of a pulley block, look at his tackle carefully. He is probably using a Weston pulley block.

This consists of two pulleys A and B of slightly different radii (see Fig. 129), which revolve together. They are fitted with notches to prevent the chain, which passes round them, from slipping.

At its lower end the chain passes round a pulley C to which is attached the load W. The chain is endless, and as it winds up on the larger pulley B it unwinds on the smaller A. Consequently

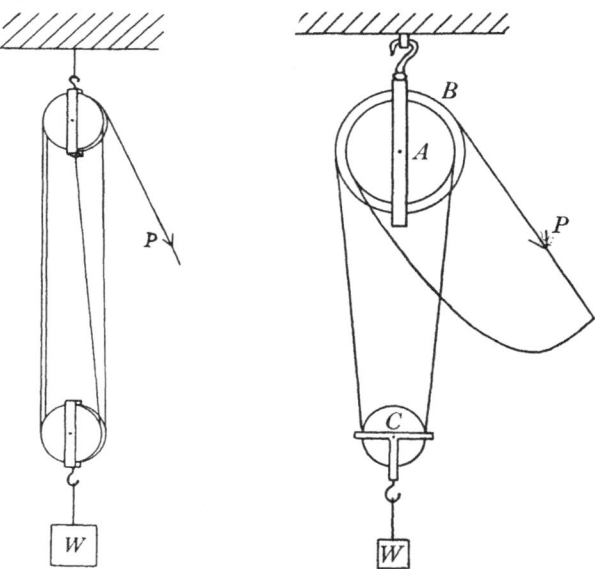

Fig. 128   Fig. 129. Weston pulley tackle.

the load rises very slowly, and the mechanical advantage is high.

Let us suppose that there are 11 teeth on A and 12 on B. The distance between successive teeth is equal to two links. After one complete revolution $(2 \times 12) = 24$ links have been wound on to B, and $(2 \times 11) = 22$ links have been wound off on A. Thus the portion of chain passing round C has been shortened by two links. The load, however, has risen a distance equal to only one link, since the chain has two sides, each of which is shortened by one link.

Thus the load has been raised one link while the effort has moved a distance of 24 links.

$$\therefore \text{ Velocity ratio} = \frac{24}{\dfrac{24-22}{2}} = \frac{24}{1} = 24.$$

If the machine were 100 per cent. efficient the mechanical advantage would also be 24. In practice, the efficiency is usually considerably less than 50 per cent. and the mechanical advantage is therefore less than 12.

Overhauling. It is sometimes useful to have a machine with an efficiency of less than 50 per cent., since such a machine will not overhaul, i.e. the load will not descend when the effort is removed. For when the efficiency is less than 50 per cent., more than half the effort is used to overcome friction, and less than half the effort to support the load. Hence the friction alone is more than sufficient to support the load when the effort is removed.

Cranes.

One of the chief uses of pulley systems is in the hoisting gear of cranes. Cranes are hoisters *par excellence*. They are called by such names as Mammoth, Titan, Hercules and Goliath according to their kind. With their help the engineer not only lifts enormous weights, but deposits them where he pleases.

Fig. 130 shows one of the pulley blocks, containing 8 pulleys side by side, used in a crane capable of lifting 200 tons. With two such pulley blocks a velocity ratio of 16 may be obtained, and consequently an ideal mechanical advantage of 16. In practice, of course, the mechanical advantage is less than 16 and an effort of more than $\frac{200}{16} = 12\frac{1}{2}$ tons is required to raise a load of 200 tons.

The hoisting gear of all cranes is very similar, but the structure from which the hoisting gear hangs is built in different forms according to the class of work the crane is required to do. For the sake of general interest we shall consider briefly the different forms of structure.

Cranes may be divided into two main classes:

1. Jib cranes.
2. Bridge cranes.

As its name implies, the most prominent feature of a jib crane is the jib, which is the long lattice girder (usually thicker in the

By courtesy of Sir William Arrol and Co., Ltd.

Fig. 130. The snatch-block of a 200-ton crane.

middle than at the ends, since this is the position of greatest strain) leaning forwards from the base. At any large port, numbers of these tall and slender cranes may be seen on the quayside, ready for unloading the cargoes of ships.

By courtesy of the Mersey Docks and Harbour Board

Fig. 131. A large floating jib crane capable of raising 200 tons to a height of 170 feet.

Over a pulley at the top of the jib passes the lifting rope or chain, one end being connected to the engines. The hanging end may be connected direct to a hook, but it usually passes round a system of pulleys.

The jib can be tilted up or down in order to alter the reach of the crane; this movement is known as derricking or luffing. It is usually capable of revolving about its base and is sometimes mounted on a carriage on rails so that it can travel backwards or forwards.

One of the largest jib cranes in England (of the kind known as Mammoth) is owned by the Mersey Docks and Harbour Board (see Fig. 131). It is a self-propelling floating crane, and lifts 200 tons to a height of 170 ft. Its jib can be derricked through about 40°.

By courtesy of Sir William Arrol and Co., Ltd.

Fig. 132. A travelling gantry crane.

The most common type of bridge crane is the travelling gantry, which may be seen at any large works. A small carriage containing the engines and hoisting gear travels on rails overhead, and from this the chain and pulley blocks are lowered. In order that this carriage or gantry may be brought to any part of the workshop, the overhead rails are arranged in the shape of a letter **H**. The carriage can move from side to side of the shed on rails corresponding to the crossbar of the letter **H**, and the crossbar can move from end to end of the shed on the side rails which run along the top of the two parallel side walls (see Fig. 132).

Another type of bridge crane is the hammerhead crane shown in Fig. 133. It consists of a horizontal head which swivels on a vertical base, rather like a hammer with a loose head. Along the horizontal head runs a carriage from which the hoisting blocks are suspended. Such a crane can lift loads up to 250 tons, and is used for lifting heavy machinery in and out of ships. Its ad-

By courtesy of Sir William Arrol and Co., Ltd.

Fig. 133. A hammerhead crane.

vantage over the jib crane in certain cases is its greater range and mobility.

Calculation of the efficiency of the pulley blocks of a crane. A crane has two pulley blocks, each containing 8 pulleys side by side, and an effort of 15 tons weight is required to lift a load of 200 tons. Find the efficiency of the pulley blocks.

Suppose each pulley block contains 8 pulleys as in Fig. 130. There are 16 lengths of chain, so that for the load to rise 1 ft. the chain must be pulled 16 ft., i.e. the velocity ratio is 16.

Further,

Load lifted $= 200$ tons wt.,

Effort exerted by engine on chain $= 15$ tons wt.

∴ Work done in raising load 1 ft. $= 200 \times 1 = 200$ ft. tons,

Work done by effort (which moves 16 ft.) $= 15 \times 16 = 240$ ft. tons.

∴ Efficiency $= \dfrac{\text{Work got out}}{\text{Work put in}}$

$= \frac{200}{240}$

$= \frac{5}{6} = 83$ per cent.

The inclined plane.

The inclined plane is a simple machine. We have reason to believe that the ancients made considerable use of it, for instance,

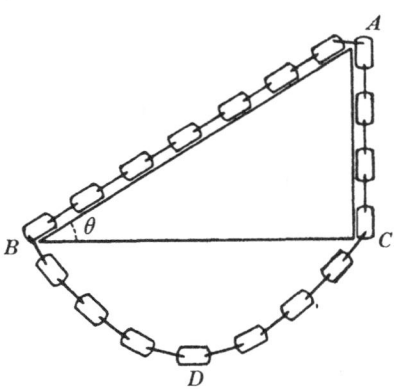

Fig. 134

in the building of the enormous Pyramids. Nowadays it may often be seen in use, when barrels are being loaded on to a brewer's dray. The barrels are rolled up inclined rails, a less exhausting process than lifting them vertically, and requiring a smaller effort, although the work done in each case is the same.

The calculation of the mechanical advantage of the inclined

plane was solved very ingeniously by Stevinus of Bruges (1548–1620).

Stevinus imagined an endless chain resting on an inclined plane (see Fig. 134). He argued that if the chain were uniform it would rest in any position. Should this not be the case, and the weight of one side cause it to move, there is no reason why it should not go on moving for ever. He assumed that such perpetual motion is impossible, and concluded therefore that the chain is balanced.

Since the portion of the chain *BDC* is symmetrical, the part *BD* balancing the part *CD*, it should make no difference to the equilibrium of the part *BAC* if the lower part of the chain were severed at *B* and *C*. Thus the weight of the vertical hanging part *AC* is balancing the weight of the part *AB* resting on the incline.

The weight of any portion of the chain will be proportional to its length.

Thus,

$$\text{Mechanical advantage} = \frac{\text{Load}}{\text{Effort}}$$
$$= \frac{\text{Weight of } AB}{\text{Weight of } AC}$$
$$= \frac{\text{Length of } AB}{\text{Length of } AC} = \frac{1}{\sin \theta}.$$

Thus if $AB = 2AC$, the mechanical advantage will be 2. What will be the velocity ratio?

Example. A barrel of beer weighing 150 lb. is to be rolled up an incline of length 10 ft. on to a cart of height 4 ft. What force up the incline is necessary?

$$\text{Mechanical advantage} = \frac{AB}{AC}.$$

From the figure,

$$\frac{150}{P} = \frac{10}{4},$$
$$P = 60 \text{ lb. wt.}$$

There is another method of solving this example. By the principle of work,

$$\text{Work done on load} = \text{Work done by effort.}$$
$$\therefore \quad 150 \times 4 = P \times 10,$$
$$\therefore \quad P = 60 \text{ lb. wt.}$$

It will be noticed that the work done on the load is obtained by multiplying the load by the vertical height raised, 4 ft., since

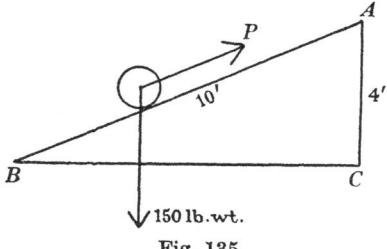

Fig. 135

the weight of the barrel acts vertically downwards. On the other hand, the effort P must be multiplied by the slant height of the plane, 10 ft., since it acts in this direction.

The screw jack.

A screw jack is often carried by motorists for raising their car in an emergency, although when compared with a hydraulic jack its operation is tedious.

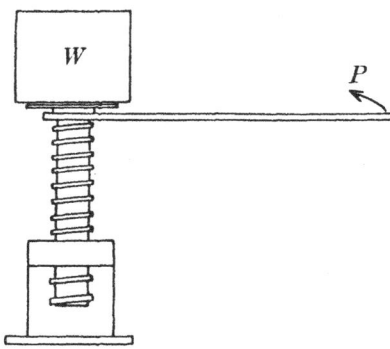

Fig. 136

It consists of a screw fitted with a heavy base, which is placed under an axle of the car. The screw is turned by means of a long lever and rises, lifting the car. The diagram shows a simplified form.

Now the screw, since it consists of a thread winding round a central shaft, is really a spiral inclined plane. Its mechanical advantage may be found by considering it in this way, having regard also to the length of the lever. Leaving this method as an exercise for the enterprising student, we shall solve the problem ourselves by means of the principle of work.

The pitch of the screw is the distance between consecutive threads measured parallel to the axis.

Suppose   Pitch of screw $= \frac{1}{4}$ in.,
             Radius of lever $= 12$ in.

When the lever is turned through one complete revolution the effort has been moved a distance $2\pi \times 12$ in., and the load has been lifted a distance equal to the pitch, $\frac{1}{4}$ in.

Let   Load $= W$ lb.,
      Effort $= P$ lb.
      Work done by effort
        in one revolution $= P \times 24\pi$ in. lb.
      Work done on load $= W \times \frac{1}{4}$ in. lb.

By Principle of Work,
      Work done on load $=$ Work done by effort.
$$\therefore \quad W \times \tfrac{1}{4} = P \times 24\pi,$$
$$\frac{W}{P} = \frac{24\pi}{\frac{1}{4}}$$
$$= 300.$$

Thus generally speaking we may say,
$$\text{Mechanical advantage} = \frac{\text{Circumference of effort arm}}{\text{Pitch of screw}}.$$

In practice the efficiency of even a well-oiled screw is small owing to friction. Thus the mechanical advantage is considerably less than the value calculated as above.

Friction.

It has been stated that the efficiency of machines is diminished by friction. Let us now consider what is meant by the term.

Friction operates whenever one body tends to slide or slides upon another. If we try to push a block of wood along a table a force of friction opposes the motion in whatever direction the block is pushed.

No surface is absolutely smooth, and even the most highly polished surface when viewed under the microscope is seen to be made up of tiny hollows and projections. Friction is caused by the interlocking of the minute projections of the two surfaces in contact.

If there were no friction, walking or locomotion of any kind would be impossible. In walking, the foot pushes backwards on the ground and the ground exerts an equal force forwards on the foot. (Action and reaction are equal and opposite.) If the ground is too slippery it cannot exert a force of friction big enough, and the foot skids. The friction may be increased by roughening the surface with sand or increasing the pressure between the foot and the ground. Thus a horse may be enabled to get a better grip on a slippery road by a man mounting on to its back.

A cyclist increases the force of friction on the rim of his bicycle wheel by applying his brake blocks more and more tightly until the bicycle stops. The billiard player cannot make a true shot unless there is a force of friction between the ball and the end of his cue, so he roughens this by marking it with chalk.

We are usually concerned, however, not to increase friction but to diminish it. One method is to lubricate the surfaces with oil, since the friction between a solid and a liquid is much less than that between two solids. A motor car engine will "seize" if it runs short of oil. The pistons and cylinders get so hot owing to the excessive friction that they become jammed together and the whole engine has to be taken down.

The coefficient of friction. The force of friction reaches its highest value when a body is on the point of moving.

If you hook a spring balance to the end of a block of wood resting on a table and pull with a gradually increasing force, the force of friction must always be exactly equal and opposite to your pull until the body begins to move. When the block is on the point of moving the friction is said to be *limiting*.

Consider the forces acting on the block. The horizontal pull of the spring balance P is balanced by the force of friction F; the weight W of the block is balanced by an upward force R exerted on the block by the table, and known as the *normal reaction* of the table.

Now we have stated that the force of friction increases when a block is pressed more tightly on to a table. In this case the normal reaction of the table must be equal to the weight of the block plus the extra downward force on the block; hence the normal reaction is the true measure of what we may loosely term "the pressure" between the block and the table (not using the word in its scientific sense of "force per unit area").

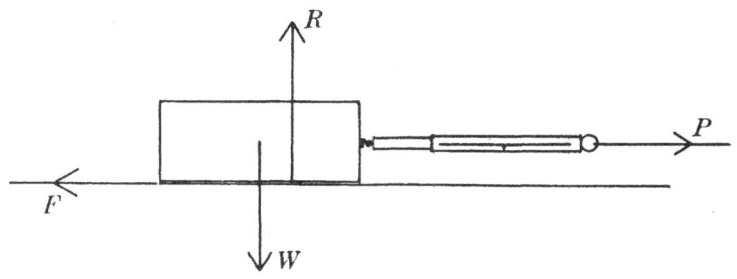

Fig. 137

It may be shown (see Experiment 4 at the end of this chapter) that the force of limiting friction F is proportional to the normal reaction R. In other words F/R is constant, whatever the value of R, for a given block and a given table. This constant is known as the **coefficient of friction** and is denoted by μ.

Thus
$$\mu = \frac{F}{R}.$$

Rolling. So far we have considered sliding only. Most vehicles, however, run on wheels, which do not slide along the ground, but roll. Moreover, the wheel is usually mounted on ball bearings, so that it rolls rather than slides on its axle (see Fig. 138).

Wherein lies the advantage of rolling over sliding? The fact is that in rolling the force of friction between the wheel and the ground has no tendency to hinder motion so long as it is large enough to prevent the wheel from sliding or "skidding". The lowest point of a wheel in contact with the ground is momentarily at rest, and does not move against the force of friction, so that no work is done against friction. In the case of the driving wheels of a vehicle it is actually desirable that the force of friction should be as *large* as possible. For a loco-motive cannot exert a greater force on the train than is provided by the friction between the driving wheels and the rails. If the rails are too smooth, the driving wheels spin round without moving the train. To prevent this, the force of friction between the rails and the driving

By courtesy of the Skefko Ball Bearing Co.

Fig. 138. Ball and roller bearings.

wheels must be acting in the same direction as the engine is moving (see Fig. 139).

The reverse applies in the case of the bogey wheels of the locomotive (those wheels to which no driving force is applied, but which are used

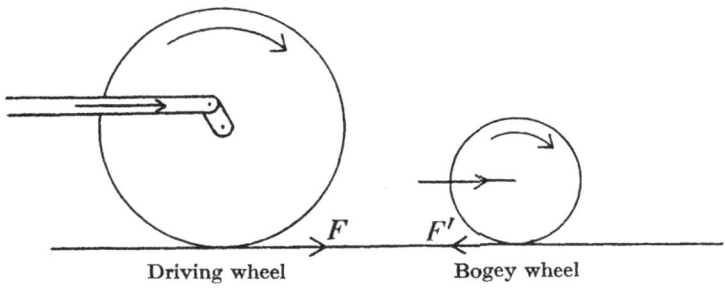

Driving wheel   Bogey wheel

Fig. 139

to bear the weight). If there were no friction between the rails and these wheels, they would slide forward without turning. Consequently the force of friction on these wheels must be acting in the opposite direction to that in which the engine is moving.

In actual practice there is always a certain amount of sliding when a wheel rolls on the ground, owing to the fact that the wheel is somewhat flattened or deformed where it makes contact with the ground. Thus in Fig. 140 some sliding must take place between AB and the ground as the wheel rolls. The more the wheel is flattened, the more important will this sliding be. Every motorist knows that his tyres wear out far more quickly if he keeps them soft instead of inflated to the correct pressure.

Experiment 1. *To find the mechanical advantage, velocity ratio and efficiency of a system of pulleys.* Set up the pulley system, hang from it a convenient load W, and adjust the effort P so that it descends slowly with a steady velocity when given a start. Hence calculate the mechanical advantage, W/P.

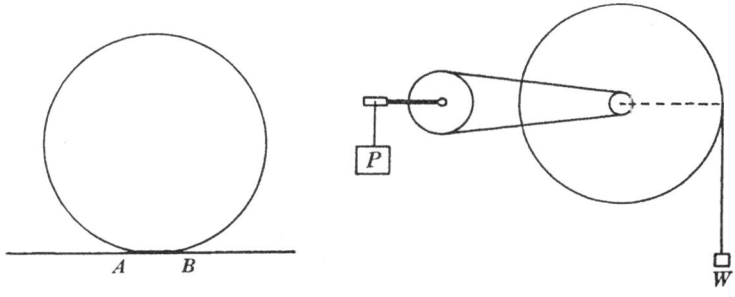

Fig. 140   Fig. 141

Move the effort through a convenient measured distance and measure the corresponding distance moved through by the load. Hence calculate the velocity ratio.

Then   Efficiency $= \dfrac{\text{Mechanical advantage}}{\text{Velocity ratio}}$ (see p. 158).

Experiment 2. *To find the mechanical advantage, velocity ratio and efficiency of a bicycle.* Find the velocity ratio first by turning the pedals forward through one complete revolution and measuring the distance moved by the bicycle. The distance moved by the pedal (on which the effort is applied) is the circumference of a circle with radius equal to the length of the crank.

$\therefore$ Velocity ratio $= \dfrac{\text{Distance moved by pedal}}{\text{Distance moved by bicycle}}.$

Suspend the bicycle so that its wheels are off the ground, and turn the pedals so that the cranks are horizontal. Hang a load W (1 lb., say) from the point on the rim of the back wheel as in Fig. 141, and balance this with an effort P hung from the forward pedal. Then the mechanical advantage $= W/P$.

Calculate the efficiency as in the previous experiment.

If the bicycle has three gears repeat the experiment in each gear.

Experiment 3. To find the mechanical advantage and velocity ratio of an inclined plane.

(*a*) Measure the force required to cause a heavy roller to move steadily up an inclined plane when given a start (Fig. 142).

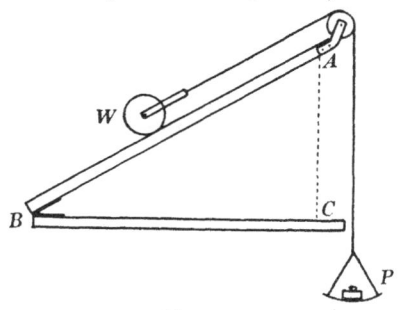

Fig. 142

Calculate,

$$\text{Mechanical advantage} = \frac{\text{Weight of roller}}{\text{Weight of scale pan} + \text{contents}}$$

and   Velocity ratio $= \dfrac{AB}{AC}.$ Why?

(*b*) Now remove some of the weights from the scale pan until the roller moves steadily down the plane when given a start. Take the average values of P when the roller moves up and down the plane, respectively, and calculate W/P. This value should equal the velocity ratio. Why?

Repeat the experiment with different inclinations of the plane, if time.

Experiment 4. To show that the force of sliding friction is pro-portional to the normal reaction (or total "pressure") between the

surfaces in contact and hence to find the coefficient of friction.
Obtain a wooden block as smooth and clean as possible, and fix
a hook in one end. Find its weight by means of a spring balance.

Place it on a smooth clean wooden board, attach a spring
balance to the hook and pull horizontally on the spring balance,
noting the reading when the block is just about to move. This
gives the limiting friction. (When the block is sliding the friction
is slightly less.)

Repeat the experiment several times, but be careful to put the
block in the same position each time; the force of friction will
probably differ slightly in different places on the board. Take the
average of these readings but record them all. The normal reaction between the block and the board is equal to the weight of
the block.

Now increase the normal reaction between the block and the
board by placing different weights on the block. Repeat the
experiment in each case.

Enter your results in two columns of a table, and on squared
paper plot force of friction against normal reaction.

The points on the graph will be found to be roughly on a
straight line. What does this prove?

Work out the coefficient of friction from each pair of readings
and insert the values in a third column in your table.

N.B. $\text{Coefficient of friction} = \dfrac{\text{Friction}}{\text{Normal reaction}}.$

*Experiment 5. Obtain a series of pairs of values of load and ex-
tension for a spiral spring and plot them in a graph. Hence find
the work done in stretching the spring (a) 5 cm., (b) 10 cm.*

If the graph is a straight line, what does this prove?

N.B.

Work done in stretching spring = Average load × extension.
$$= \tfrac{1}{2} \text{ final load} \times \text{extension.}$$

SUMMARY

The following terms are used in connection with
machines:

$$\text{Mechanical advantage} = \frac{\text{Load}}{\text{Effort}} = \frac{W}{P}.$$

$$\text{Velocity ratio} \quad = \frac{\text{Distance moved by effort}}{\text{Distance moved by load}} \cdot$$

$$\text{Efficiency} \quad = \frac{\text{Work got out}}{\text{Work put in}}$$

$$= \frac{\text{Mechanical advantage}}{\text{Velocity ratio}} \cdot$$

Work = Force × distance moved in the direction of the force (ft. lb.).

All machines are subject to the Principle of Work: "A machine cannot give out more work than is put into it".

Pulley systems, the inclined plane and the screw are simple machines of fundamental importance to which the above theory has been applied in this chapter.

Friction is the force tending to prevent one body sliding on another.

$$\text{Coefficient of friction} = \frac{\text{Force of limiting friction}}{\text{Normal reaction}} \cdot$$

Normal reaction is the force at right angles to the surfaces, representing the "pressure" between them.

QUESTIONS

1. What is the purpose of a 3-speed gear on a bicycle? Describe carefully the effect on the velocity ratio and the mechanical advantage of the machine when a change is made from (a) bottom to middle, (b) top to middle gear. Would you expect the efficiency to alter?

2. Draw a pulley system with a velocity ratio of (a) 6, (b) 5.

3. Define mechanical advantage, velocity ratio, and efficiency. Prove that if the velocity ratio and mechanical advantage of a machine were equal, its efficiency would be 100 per cent.

4. A pulley system is used to raise a load of 1 cwt. and then later a load of 1 ton. Do you think that (a) the mechanical advantage, (b) the velocity ratio, (c) the efficiency will differ in the two cases? Give your reasons.

5. The cranks of a bicycle are 7 in. long, and the diameter of the back wheel is 28 in. The wheels over which the chain passes have 48 and 24 teeth respectively. Calculate the velocity ratio of the machine. The mechanical advantage is found to be $\frac{1}{8}$; work out the efficiency.

6. In a Weston pulley block the larger pulley has 13 teeth, and the smaller 12 teeth. What is its velocity ratio? If a load of 56 lb. wt. is raised by an effort of 9 lb., what is the efficiency of the system?

7. What length of incline will be necessary to raise a garden roller weighing 400 lb. to a height of 3 ft. if the man pulling on it can exert a force of only 150 lb.?

8. A screw press has a screw with a pitch of $\frac{1}{2}$ in. and the handles used for turning it are 9 in. long. What is its velocity ratio? If its

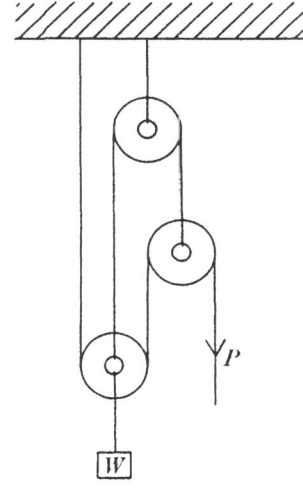

Fig. 143

efficiency is 30 per cent., what effort will be necessary to exert a pressure of 1 ton?

9. The velocity ratio of a Weston pulley block is 16. When used to raise varying loads it gave the following results:

Load (lb.)	0	14	28	42	56	70	84
Effort (lb.)	2	4	$6\frac{1}{4}$	$8\frac{3}{4}$	11	$13\frac{1}{2}$	$15\frac{1}{2}$

Plot a graph of load against effort.

Work out the efficiency of the machine for each of the loads, and plot another graph of load and efficiency. What can be deduced from the shape of the two graphs?

10. Fig. 143 represents the Barton system. What is the velocity ratio? If the efficiency is 80 per cent., what effort will be required to lift a load of 100 lb.?

11. A wireless pole has to be raised from the ground. In the position shown in Fig. 144, what force applied to the best advantage is necessary

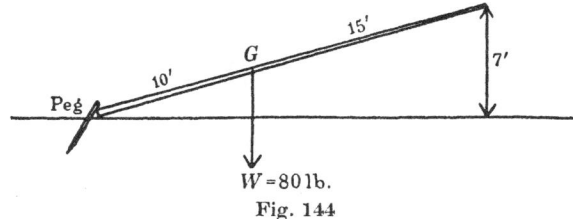

Fig. 144

to lift the pole? How much work must be done altogether in raising the pole to its vertical position?

12. A locomotive has six coupled driving wheels and the total weight on these three pairs of wheels is 60 tons; if the coefficient of

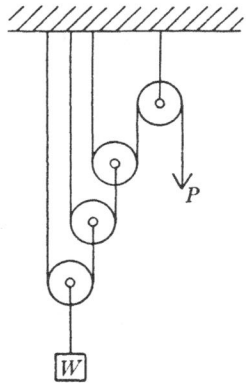

Fig. 145

friction between the wheels and the rails is 0·2, what is the greatest pull which the engine can exert without causing wheel slip? What difference would it make if only a single pair of driving wheels were connected?

13. What is the velocity ratio of the system of pulleys shown in Fig. 145? If each of the pulleys weighs ½ lb., what effort will be re-

quired to raise a load of 14 lb.? Hence find the mechanical advantage and efficiency for this load. [Hint. Find the tension in each string by considering the equilibrium of each pulley in turn.]

School Certificate Questions

14. Distinguish between the mechanical advantage and the efficiency of a machine.

A pulley system consists of five pulleys arranged in two blocks, the lower movable block containing two pulleys, the same string passing round all the pulleys. A force of 4 lb. just supports a weight of 18 lb. Sketch the system and calculate its efficiency.

15. A motor car jack has a screw of pitch 0·6 in. and a handle measured from the axis of the screw to the end of the handle, 15 in. long. A force of 25 lb. wt. applied to the end of the handle will lift a weight of 1250 lb. Calculate the mechanical advantage, the velocity ratio, and the efficiency of the jack.

16. Four men raise a ship's anchor weighing 1 ton by means of a capstan 12 in. in diameter, the levers of which are 6 ft. long measured from the axis of the capstan. If the efficiency of the machine is 80 per cent., what force must be exerted by each man?

Chapter VIII

POWER AND ENERGY

Horse-power.

The most important thing about a machine is not how much work it can do, but how fast it can do it. A disreputable "old crock" of a car can get from London to Brighton given long enough, and does just as much effective work as the limousine which completes the journey in an hour and a half. Yet the limousine is more powerful, and can therefore do the work faster.

Before the advent of steam engines horses were used to turn treadmills by walking round in a circle at the end of a long bar. When James Watt (1736–1819) began to manufacture steam engines he had to tell his customers how many horses they would replace. He therefore devised an experiment to find the rate at which a horse could work.

He chose heavy dray horses from Messrs Barclay and Perkins' brewery, London, and set them to pull up a weight of 100 lb. from the bottom of a deep well. He found that they could walk comfortably at $2\frac{1}{2}$ m.p.h. when raising this load.

Now,
$$2\frac{1}{2} \text{ m.p.h.} = \frac{2\frac{1}{2} \times 1760 \times 3}{60 \times 60} \text{ ft. per sec.}$$

$$= 3\frac{2}{3} \text{ ft. per sec.}$$

Work done in raising 100 lb. through $3\frac{2}{3}$ ft.

$$= 100 \times 3\frac{2}{3}$$

$$= 366\frac{2}{3} \text{ ft. lb.}$$

Thus a horse could do $366\frac{2}{3}$ ft. lb. per sec. To be on the safe side, Watt allowed an extra 50 per cent. for work expended in overcoming friction. He was a cautious man and did not wish to over-estimate the power of his steam engine. He took **1 horse-power** as **550 ft. lb. per sec.** It has since been found that the average horse can work continuously at only about three-quarters of this rate. However, Watt's unit is still used for rating the power of engines.

Brake horse-power.

It is clearly impracticable for a manufacturer to have a deep well in his workshop to test the power of a machine. Instead, he runs it on a bench (in the case of a motor car engine) and applies a brake. The brake is of such a form that the work done against it can be determined. The horse-power of the engine obtained in this way is known as brake horse-power.

We will describe a simple friction brake.

A belt passes over a flywheel driven by the engine, and it is

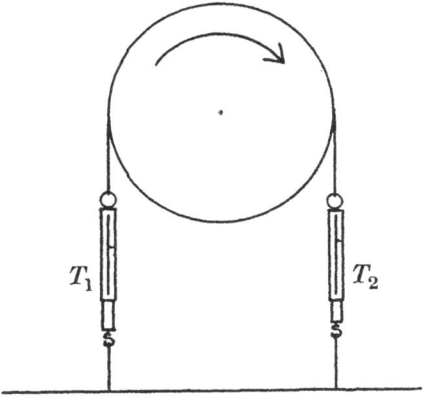

Fig. 146

kept tight by attaching its ends to two spring balances (tensions T_1 and T_2). The friction between the belt and the flywheel is the braking force.

With the wheel rotating as in Fig. 146, T_1 is clearly greater than T_2, because the wheel is trying to drag the belt with it: T_2 helps the wheel in its motion, and T_1 hinders. Thus the effective force to be overcome is $T_1 - T_2$.

The faster the engine revolves, the greater the horse-power it is developing, for it is turning the wheel through a greater distance against the same friction. The total distance moved per sec. is equal to the circumference of the wheel multiplied by the number of revolutions per sec. (recorded by a counter). Hence the work done by the engine per sec. and its horse-power can be found.

The actual method of working out will be shown in the following experiment.

An experiment to find the horse-power of a boy. An apparatus similar to the one shown in Fig. 146 is used. There is a handle to the wheel and the boy turns the handle as fast as he can for 1 minute, the number of revolutions made in this time being counted. He would not be able to keep up this rate for long, so that the answer will represent the short bursts of horse-power he can exert with his arms. The radius of the wheel must be found, and the average values of the spring balance readings T_1 and T_2 noted while the wheel is being turned.

    Readings (simplified).

	Radius of wheel	$= 1$ ft.
	Number of revs. in 1 min.	$= 35$.
$\therefore$	Total distance turned	$= 2 \times 1 \times 35$ ft.
		$= 2 \times \frac{22}{7} \times 35$ ft.
		$= 220$ ft.
	Average value of T_1	$= 28$ lb.
	Average value of T_2	$= 7$ lb.
$\therefore$	Force of friction	$= T_1 - T_2$
		$= 21$ lb.
$\therefore$	Work done per min.	$= 21 \times 220$ ft. lb.
But	1 horse-power	$= 550$ ft. lb. per sec.
		$= 33,000$ ft. lb. per min.
$\therefore$	Horse-power	$= \dfrac{21 \times 220}{33,000} = 0.14$.

Indicated horse-power.

The brake horse-power of an engine represents the rate at which it actually does work; the indicated horse-power is a calculation, from the size of pistons and pressure of steam or petrol vapour in the cylinders, of what the rate of working ought to be. The brake horse-power is always less than the indicated horse-power owing to internal friction in the engine.

How the horse-power of a machine changes.

The horse-power a car develops depends on the speed at which its engine is revolving. A light, small horse-powered car can climb really steep hills in top gear if it is travelling fast enough when it begins the climb. If the hill is too steep and the engine begins to slow down, the horse-power drops and the driver has to change into a lower gear in order to speed up his engine again.

Thus in the case of a car listed as 9–20 H.P., 20 H.P. is the horse-power the engine must develop when running at full speed on a test bench. The 9 H.P. is the Treasury rating for taxation purposes. It represents no real scientific quantity, but is calculated from the bore of the cylinders.

The horse-power of an engine increases steadily as its revolutions increase only up to a certain point. In the case of a motor car, this is the point where the driver's foot is right down on the accelerator pedal—when he cannot further increase the rate at which the petrol flows to the engine. The petrol which enters the cylinders when the throttle is fully open is not sufficient to exert any further pressure on the pistons, since they are already moving so rapidly.

Racing car and aeroplane engines are fitted with super-chargers, which blow the petrol vapour into the cylinders under pressure, thereby considerably increasing the maximum horse-power. Any car could be "hotted up" with this device.

In the same way the maximum horse-power of a railway locomotive depends on its steam-raising capacity; hence the importance of a competent stoker.

Again, a boy can walk upstairs or run upstairs. In the latter case he is developing a greater horse-power than in the former. The human body acts like an internal combustion engine, the fuel being digested food. In order to transform this fuel into muscular energy, it must be "burnt" by chemical combination with oxygen. The capacity of the lungs governs the rate at which the body can take in oxygen, and the efficiency of the heart governs the speed at which the oxygen and digested food can be delivered to the muscles. Hence the maximum horse-power of a boy depends primarily on the condition of his heart and lungs.

Example. To calculate the resistance of road and air on a car. Suppose a car developing 20 H.P. can travel at a maximum speed of 45 m.p.h. on the level.

$$\begin{aligned}
\text{Power of car} \quad &= 20 \text{ H.P.} \\
&= 20 \times 550 \text{ ft. lb. per sec.} \\
&= 11,000 \text{ ft. lb. per sec.}
\end{aligned}$$

Let   Resistance of road and air $= R$ lb. wt.

       Speed of car $= 45$ m.p.h.

$$= 45 \times \frac{1760 \times 3}{60 \times 60} \text{ ft. per sec.}$$

$$= 66 \text{ ft. per sec.}$$

Then   Work done per sec. against

         resistance $= \text{Force} \times \text{distance}$

$$= R \times 66 \text{ ft. lb.}$$

But   Work done per sec. by engine $= 11{,}000$ ft. lb.

$$\therefore \quad 66R = 11{,}000$$

$$\therefore \quad R = \frac{11{,}000}{66} = 167 \text{ lb.}$$

(to the nearest lb.)

Potential energy.

If we lift a 3 lb. weight through a vertical height of 4 ft. we have done 12 ft. lb. of work. We have done work on the weight and it is now in a position to do work for us. There is stored up in it 12 ft. lb. of work. A grandfather clock is driven in this way. A weight is wound up, perhaps once a week, and slowly descends, driving the clock.

The capacity of a body for doing work is called its **Energy.** A body which can do work by virtue of its position, such as the wound-up weight in a grandfather clock, is said to have **Potential Energy.**

A wound-up clock spring also has potential energy, because it can do work as it uncoils. A bent bow has its own potential energy which it can impart to an arrow. The chemical energy of coal or oil, which can be used to drive an engine, is really a form of potential energy.

Example. The potential energy of some of the water at the top of the Niagara falls is absorbed by means of turbines and converted into electricity. The falls are about 160 ft. high, and roughly 660,000 tons of water pour over the top per min. We will calculate the total available horse-power of the falls if the potential energy of all the water could be utilised.

Weight of water falling

     per min. $= 660{,}000$ tons

$$= 660{,}000 \times 2240 \text{ lb.}$$

But   1 horse-power $= 550$ ft. lb. per sec.

$= 550 \times 60 = 33,000$ ft. lb. per min.

$\therefore$ Available horse-power $= \dfrac{660,000 \times 2240 \times 160}{33,000}$

$= 7,168,000$

$=$ approx. $7,000,000$.

Kinetic energy.

We have seen that a machine cannot produce more work than is put into it. It cannot create energy. Neither can energy be destroyed. When energy apparently disappears it has merely been changed into some other form. This is one of the most important laws of all Physical Science, and is known as the **Principle of the Conservation of Energy.**

Consider a body of weight 3 lb. falling freely through a height of 4 ft. As it falls it is losing potential energy because it has not so far to fall. However it is gathering speed, and the faster it moves the more work it could do if it struck anything. Its potential energy is being converted into **Kinetic Energy**—energy by virtue of motion.

A good example of a constant interchange of kinetic and potential energy is the switchback. At the bottom of each dip the car has a maximum of kinetic energy and at the top of each rise most of the kinetic energy has been converted into potential energy. If there were no friction, a car would be able to rise to the same height from which it had dropped. That part of the car's energy which overcomes friction is changed into heat between the car and the rails, and in this form is dissipated into the atmosphere.

The pile driver.

During the building of breakwaters and bridges it is often necessary to drive large posts of timber or reinforced concrete, called piles, into the bed of the sea or stream. The instrument used for doing this is known as a pile driver and is a good example of the kinetic energy of a moving body being used to do work. It consists of a tall tower of scaffolding up which a heavy weight is drawn and allowed to fall on to the head of the pile (see Fig. 147). The weight is repeatedly drawn up and allowed to fall until the pile has been driven in.

Fig. 147. A pile driver driving a concrete pile into the ground during the preparation of the foundations of the Stratford-on-Avon Shakespeare Memorial Theatre.

Calculation of the force exerted by a pile driver.

Suppose   Falling weight = 500 lb.

Distance fallen = 4 ft.

∴ Kinetic energy of falling weight just before the blow

= Loss of potential energy

$= 500 \times 4$ ft. lb.

$= 2000$ ft. lb.

Let us suppose that the pile is driven in 1 in., i.e. $\frac{1}{12}$ ft., at each blow.

Let R lb. = Average resistance of ground.

Work done against resistance = Force × distance

$= R \times \frac{1}{12}$ ft. lb.

But the kinetic energy of the falling weight disappears, and we will assume all of it has gone to overcome the resistance of the ground.

Then   Kinetic energy lost = Work done.

$$2000 = R \times \tfrac{1}{12}.$$
$$\therefore \quad R = 24{,}000 \text{ lb.}$$

Thus the blow delivered by the pile driver is equivalent to a steady force of 24,000 lb. This example shows that enormous forces can be produced when a body in which kinetic energy is stored is brought suddenly to rest. In practice not all of the energy is used in overcoming the resistance of the ground; some is wasted in distorting the end of the pile. The effect is to reduce the force of the blow.

The flywheel.

A flywheel is a massive wheel mounted on ball bearings so that it revolves with very little friction. Owing to its great mass a considerable amount of work is required to set it revolving at speed.

The work done in starting is stored up in it as kinetic energy, and it can do an equivalent amount of work before being brought to rest.

Flywheels are used in engines, for instance in a gas engine or a steam roller, to store up energy during the working part of the stroke of the piston, and give it out during the non-working part. Thus the machine is made to run steadily and not in a series of jerks.

A punching machine capable of making a hole in a plate of cold steel an inch thick stores up energy in a massive flywheel, and then brings it all to bear on the plate to be punched. As the wheel is brought to rest its kinetic energy is transformed into work. The work done will depend on the force necessary to penetrate the steel, and the thickness of the plate (i.e. the distance this force has to travel).

Units of work and power on the metric system.

Only the British units of work and power, the ft. lb. and the horse-power respectively, have so far been mentioned, and a very wrong impression of the subject of mechanics would be given if no mention were made of the corresponding units on the metric system. For the metric system is, so to speak, the official system of units. It is international in scope, and is used in all branches of science.

The unit of work on the metric system is the *joule*. This represents approximately the work required to raise 1 kilogram a vertical distance of 10 cm. (The precise definition is given on p. 258.) The metric unit of power is the *watt*, and is defined as a rate of working of 1 joule per sec. A *kilowatt*, a larger and therefore sometimes more convenient unit, is 1000 watts. The kilowatt is roughly equivalent to $1\frac{1}{3}$ H.P., or, more accurately, 746 watts = 1 H.P.

A 60 watt electric lamp is one in which work must be done at the rate of 60 watts if it is to give the illumination for which it is designed. A 100 watt lamp, of course, requires a greater rate of working, and gives a brighter light. If ten 100 watt lamps are burning together, they are using up energy (or work) at the rate of a kilowatt, and if they burn for an hour they use up a definite amount of work, known as 1 kilowatt hour. By the words "use up" is meant the conversion of electrical energy into light and heat.

Electricity companies sell their electrical energy at so much (perhaps 3*d*. or 6*d*.) per "unit", the unit being the kilowatt hour. Thus the consumer buys electrical energy in terms of the mechanical work to which it is equivalent.

SUMMARY

The rate at which a machine does work is measured in horse-power.

1 horse-power = 550 ft. lb. per sec.

Energy is "stored-up" work, and the energy possessed by a body represents its capacity to do work. Energy by virtue of position is known as **Potential Energy,** and energy by virtue of motion is known as **Kinetic Energy.** Both are measured in the unit of work, the ft. lb. (or the joule).

Principle of the Conservation of Energy: Energy can never be created or destroyed. It can, however, be changed into some other form.

QUESTIONS

1. Define horse-power.
An engine raises 10 tons of coal per min. from the bottom of a shaft 900 ft. deep. What horse-power is it developing?

2. A locomotive travelling at 60 m.p.h. is exerting a pull of 3 tons wt. on the train. What horse-power is it developing?

3. A car weighing 1200 lb. ascends a hill with a gradient of 1 in 6 at a speed of 15 m.p.h., i.e. 22 ft. per sec. Find how much work is done per sec. (a) in raising the car, (b) in overcoming a total force of friction and air resistance equal to 50 lb. wt.

4. Explain what is meant by the term "energy". What is the difference between potential and kinetic energy? A pendulum is set swinging. Describe how its energy changes in each swing. When it comes to rest, where do you think its energy has gone?

5. Explain why a falling cricket ball can exert a greater force than its weight. Why does a cricketer draw back his hands while catching a fast ball?

6. A fire engine can throw 390 gallons of water per min. to a height of 120 ft. What horse-power does this represent? (1 gallon of water weighs 10 lb.)

7. A "Bristol Jupiter" aero-engine develops 500 H.P. What force is being exerted by the propeller when the aeroplane is travelling at 120 m.p.h.?

8. A steam shovel lifts 1 ton of earth to a height of 12 ft. in 4 sec. If an engine working at 20 H.P. is used, find how much work is wasted per sec., and the efficiency of the steam shovel.

9. At Fully power station in Switzerland the turbines generate 3000 H.P. The head of water used is 5320 ft. Assuming an efficiency of 80 per cent., calculate the flow of water in tons per min.

10. An arrow is discharged from a bow at an angle of 45° with the ground; trace all the transformations of energy which go on from the moment when the archer begins operations to the moment when the arrow strikes the ground.

11. A chain weighing 3 lb. per ft. length hangs down a shaft and is wound up on to a drum of radius 1 ft. Show that if it is wound up at a constant speed of 70 revs. per min. the horse-power needed becomes less as it is wound up. Find the total work done and the *average* horse-power if the chain is 400 ft. long.

School Certificate Questions

12. Explain the terms force, work and energy. Illustrate by familiar examples the principle of the Conservation of Energy.

13. Explain why the blow of a light hammer on the head of a nail can exert a considerably greater force than a much heavier mass laid on the head of the nail.

14. A train and engine weighing together 100 tons are kept moving with a uniform speed of 30 m.p.h. on a level track. The resistance to motion due to air and friction is 40 lb. per ton. What is the horse-power developed by the engine?

15. A motor car of mass 25 cwt. climbs a gradient of 1 in 20 at 30 m.p.h. Neglecting friction, find the horse-power.

16. What is the relation between the horse-power and the watt?

Find the average horse-power in each case of a man who (a) runs up a flight of steps 10 ft. high in 2 sec., (b) climbs a mountain 3000 ft. high in 3 hours. The weight of the man is 150 lb.

17. The crew of an eight-oared boat makes 35 strokes per minute. The pull in each stroke is for a distance of 5 ft. and the average pull per man during each stroke is 70 lb. wt. How much work does each man do per minute, and what is the horse-power of the crew?

18. A 56 lb. weight is dropped from a height of 25 ft. on to soft ground which it penetrates 6 inches: (a) What available energy is in the weight at the start? (b) What is the average force of resistance of the ground?

19. A pile driver consists of a block weighing 2 cwt. falling 6 ft. on the head of the pile. Calculate in ft. lb. the kinetic energy of the block when it strikes the pile. If all this energy is expended in driving the pile 1 in., how much work is done, and what is the average resistance to the motion of the pile, in lb. wt.? Is it possible for the whole of the kinetic energy of the block to be thus expended? If not, what becomes of the balance?

Chapter IX

THE PARALLELOGRAM LAW

The Parallelogram of Velocities.

Suppose a ship is steaming due N. at 20 ft. per sec., and a man walks along the deck, also due N., at 6 ft. per sec. He is clearly moving with respect to the sea at a speed or velocity of $20 + 6 = 26$ ft. per sec. This is called his *resultant velocity* (the combined result of the velocities).

If he walks due S., in the opposite direction to the movement of the ship, his resultant velocity is $20 - 6 = 14$ ft. per sec. But

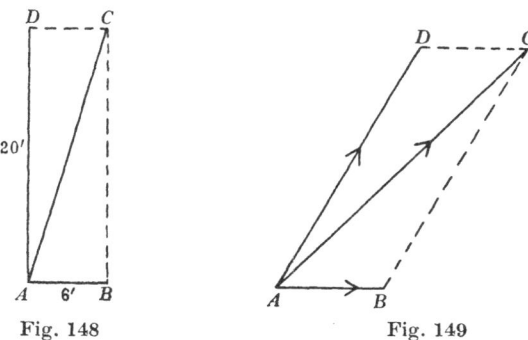

Fig. 148   Fig. 149

what is his resultant velocity if he walks due E. at right angles to the ship's course?

Look at Fig. 148. Suppose the man starts at *A*. In one sec. he walks 6 ft. E., *AB*, and is carried 20 ft. N. by the ship, *AD*. At the end of 1 sec. he will therefore be at *C* and the distance he has travelled in 1 sec. over the sea is *AC*. Thus his resultant velocity is represented in magnitude and direction by the diagonal *AC* of the rectangle *ABCD*.

If the man walks in a direction not at right angles to that of the ship, a similar method may be used to find his resultant velocity, but *ABCD* will be a parallelogram instead of a rectangle.

AB and *AD* (see Fig. 149) are drawn in the direction of the two velocities and proportional in length to their magnitude. Then the parallelogram *ABCD* is completed, and the diagonal *AC* is drawn. This represents the resultant in magnitude and direction.

We may state the principle in its general form as follows:

If two inclined velocities are represented in magnitude and direction by the adjacent sides, AB, AD, of a parallelogram ABCD, then their resultant is represented in magnitude and direction by the diagonal AC.

Vectors.

We have seen that if two velocities are in the same direction their resultant may be obtained by simple addition, and if they are in opposite directions, by subtraction. But if they are in different directions their resultant must be found by the parallelogram construction.

Quantities which have direction as well as magnitude are known as *vectors.* Another vector quantity besides velocity is force. To describe a force completely we must not only state how big it is, but also in which direction it acts. A force, and indeed any vector quantity, can be most conveniently represented on paper by a straight line, the length of the line representing the magnitude of the force, and the direction of the line representing the direction of the force. We have, of course, already used this method of representing velocity.

Most quantities, for instance a man's wealth or the number of oranges in a box, are not vectors; they have magnitude but no direction, and are sufficiently described by a number and the appropriate unit. They are called *scalar* quantities.

The Parallelogram of Forces.

Suppose a sledge is pulled by two ropes in different directions as in Fig. 150. The sledge will move in some intermediate direction. It is clear that if a third rope were pulled in this intermediate direction with the requisite force it would have the same effect as the pull of the first two ropes. Speaking generally, we may say that any two forces at an angle may be represented by a single equivalent force, which is known as the *resultant.*

Now force is a vector quantity like velocity, and the resultant of two inclined forces may be found by the same method as that used to find the resultant of two velocities—the parallelogram construction.

To find the resultant of two inclined forces acting on a body, draw two adjacent sides of a parallelogram in the direction of the forces, and proportional in length to the magnitude of the forces. Then the diagonal will represent the resultant in magnitude and direction.

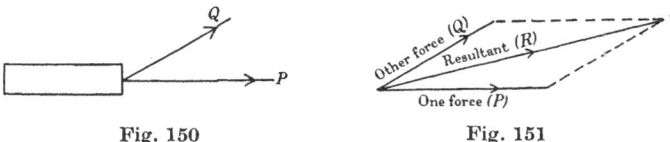

Fig. 150   Fig. 151

Do the following example by actually constructing a parallelogram:

Example. A sledge is pulled by two ropes, at an angle of 30° with each other, and the pulls are 30 lb. wt. and 20 lb. wt. Find the magnitude and direction of the resultant. Take as your scale 1 in. represents 10 lb. wt. [Answer = 37·5 lb. wt. at an angle of 12° with the 30 lb. force.]

We shall make no attempt to give a theoretical proof of this construction, but content ourselves with describing (later) an experiment whereby it can be verified in the laboratory.

Three forces in equilibrium.

The pulls P and Q of two men on a sledge can be balanced by a third pull S, and the sledge will remain stationary, i.e. in equilibrium.

We can find the necessary magnitude and direction of the pull S by means of the parallelogram of forces, since it must be exactly equal and opposite to the resultant R of P and Q (see Fig. 152).

Thus, generally speaking, we may say that three forces will be in equilibrium if one of them is equal and opposite to the resultant of the other two. If a parallelogram is drawn with adjacent sides representing two of the forces, then its diagonal (reversed) must represent the third force.

There is one other important fact which we must notice here. **If three forces are in equilibrium, their lines of action must**

meet in a point, since the resultant of any two of them must be in the same line as the third force, in order to balance it.

This fact is often useful in the solution of problems. If three forces are in equilibrium and the direction of two of them is known, the direction of the third force, such as the reaction of a hinge which may act in any direction, may be found by joining the hinge to the point of intersection of the first two forces.

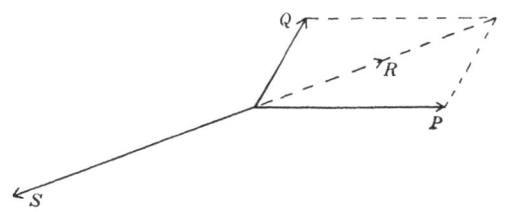

Fig. 152

An experiment to verify the parallelogram of forces. For this experiment we need to set up two inclined forces, and then find their resultant.

The inclined forces may be applied by strings passing over pulleys and carrying convenient weights, and the simplest method of finding their resultant is to balance them by a third force which will be equal and opposite to their resultant.

Set up the apparatus shown in Fig. 153. Weights in the ratio of those shown in the figure will be found to be convenient, though no such proportion is essential.

Place behind the strings a vertical drawing-board, and arrange its position so that the knot where the strings are joined is roughly in the middle of the paper. Make sure that the weights are hanging quite freely (not touching the drawing-board, for instance) and mark the positions of the strings on the paper. This may be done most accurately by placing one of the right-angle sides of a set square against the drawing-board, moving the set square up so that the other right-angle side touches the string, and making a fine mark on the paper with a sharp pencil. Make two such marks, as far apart as possible, for each of the three strings and then take down the drawing-board.

Join up the points and mark off a distance of 2 in. from the knot along the string in which the pull is $\frac{2}{10}$ lb. and 3 in. along the string in which the pull is $\frac{3}{10}$ lb. Complete the parallelogram and hence find their resultant by drawing the diagonal. If the parallelogram of forces is true, the diagonal should be 4 in. long,

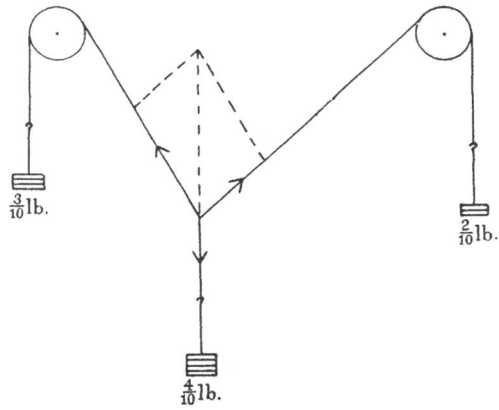

Fig. 153

and in the same direction as the string which supported the $\frac{4}{10}$ lb. wt. Measure the actual length of the diagonal and find the percentage error.

Example.

Diagonal $= 3 \cdot 9$ in.

Error $= 0 \cdot 1$ in.

$\therefore$ Percentage error $= \dfrac{0 \cdot 1}{4 \cdot 0} \times 100$

$= 2\frac{1}{2}$ per cent.

The Triangle of Forces.

In Fig. 152 the three lines representing the forces P, Q, S, which are in equilibrium, can be fitted into a triangle without changing their length or direction. It will be seen that the triangle is the same as that formed by two adjacent sides and the diagonal (the one drawn) of the parallelogram. Hence follows a corollary to the Parallelogram of Forces:

If three forces acting at a point are in equilibrium, they

can be represented in magnitude and direction by the sides of a triangle, taken in order.

The phrase "taken in order" means that the arrows, indicating the directions of the forces, must follow each other round in the same direction—either clockwise or anti-clockwise. If one of the arrows were reversed, it is clear that the forces could not be in equilibrium.

Example 1. A 3 lb. wt. is supported by two strings, inclined at 30° and 40° to the vertical respectively. Find the tension of the strings. Three forces acting through the point O are in equilibrium. Construct the triangle of forces, ABC.

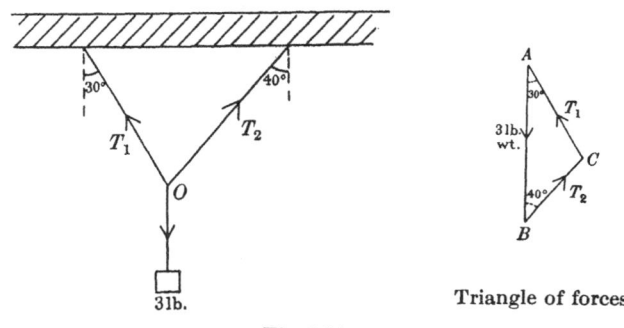

Triangle of forces

Fig. 154

Draw AB 3 in. long to represent the 3 lb. wt.

Draw AC and BC parallel to the directions of T_1 and T_2 respectively, intersecting in C.

Measure the lengths of AC and BC.

Then
$$AC = 2\cdot05 \text{ in.,}$$
$$BC = 1\cdot60 \text{ in.}$$
$$\therefore \ T_1 = 2\cdot05 \text{ lb. wt.,}$$
$$T_2 = 1\cdot60 \text{ lb. wt.}$$

Example 2. A uniform rod AB, of weight 8 lb., is hinged to a wall at A and held horizontal by a string attached to the end B and inclined at 40° to the vertical. Find the tension of the string, and the reaction at the hinge. There are three forces acting on AB, its weight, the tension of the string, T, and the reaction at the hinge, R. Since these three forces are in equilibrium, their lines of action must meet in a point. The weight of the bar acts vertically through its mid-point. Suppose its line of action intersects the string in O. Then the reaction at the hinge must act along AO.

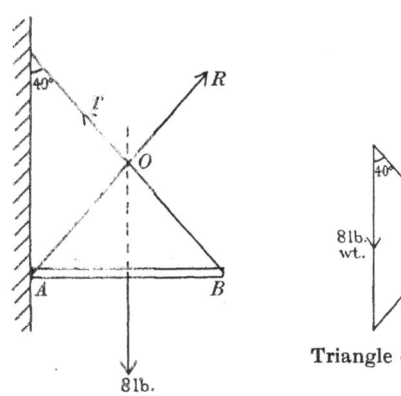

Fig. 155

Having drawn an accurate diagram of the bar and the forces acting on it (to find the direction of R), draw the triangle of forces as in Example 1. Hence find T and R.

[*Answer.* $T = 5.22$ lb. wt.; $R = 5.22$ lb. wt. at an angle of 40° to the vertical.]

Example 3. A body of weight W lb. is just on the point of slipping down an inclined plane, when the latter is tilted at an angle of λ with the horizontal. If μ is the coefficient of friction between the body and

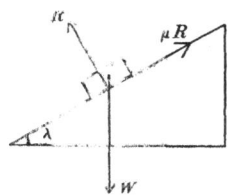

Fig. 156

Triangle of forces

the plane, show that $\mu = tan\ \lambda$. There are three forces acting on the body, its weight W, the normal reaction of the plane R, and the force of friction μR.

These three forces are in equilibrium, and hence can be represented by the sides of a triangle.

In the triangle of forces, $\quad tan\ \lambda = \dfrac{\mu R}{R} = \mu.$

[λ is called the angle of friction.]

Resolving a single force into two components.

Just as two forces can be represented by a single force, so one force can be split up into two equivalent forces. The process is known as *resolving a force into two com-ponents.* For instance, a force F (see Fig. 157) may be resolved into two components at right angles, X and Y. Draw a line AC in the direction of F and pro-portional in length to the magnitude of F, and taking it as a diagonal draw round it a parallelogram (or rectangle in this case).

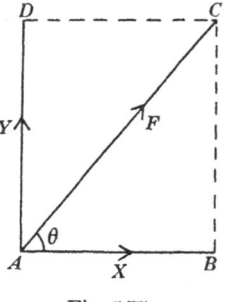

Fig. 157

The sides AB and AD represent the components of F.

It is clear from the diagram that $BC = AD$, and therefore it is only neces-sary to draw the triangle ABC instead of the whole rectangle $ABCD$. Those whose mathematical knowledge includes trigonometry will realise that $X = F \cos\theta$ and $Y = F \sin\theta$, so that the components may be calculated without drawing a triangle or rectangle to scale.

Example. A horse pulls a barge along a canal: the rope is in-clined at 30° *to the towpath. If the horse pulls with a force of* 200 *lb.*

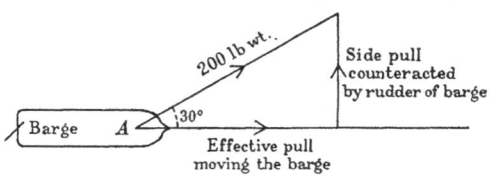

Fig. 158

wt., what is the effective pull (or component) in the direction the barge is moving? Solve this example for yourself on paper by drawing a figure similar to Fig. 158 (which is not drawn to scale). [*Answer.* 173 lb. wt.]

How an aeroplane flies.

We can now make use of the foregoing theory to examine the problem of how an aeroplane flies. When an aeroplane rises from the ground and flies in the air, it is clear that something must be supporting its weight. If it remains at a steady height, there

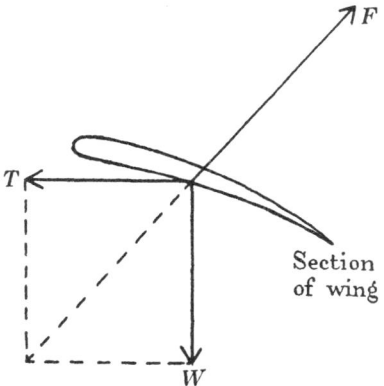

Fig. 159. The forces acting on an aeroplane in flight.

must be a vertically upward force acting on it which is equal to its weight.

The necessary force is provided by the upthrust of the wind which the aeroplane helps to make as it travels along. There is a minimum speed below which the force of the wind is insufficient. Consequently the aeroplane requires a certain amount of ground to get up speed before rising, or if launched from a ship, has to be catapulted from the deck. The early inventors did not realise this, and an aeroplane built by Langley, which was flown successfully years after his death, crashed at once when he launched it from the top of a houseboat because it did not acquire a sufficient speed.

The wings of an aeroplane are tilted up slightly from back to front, so that the wind can get under them. The result is a region of increased air pressure under the wings and one of reduced air pressure above them. The combined effect produces a wind force on the wings tilted back roughly at right angles to them.

Thus there are three forces acting on an aeroplane in flight (see Fig. 159), its weight W, the thrust forward due to the propeller T, and the wind force F. When the aeroplane is steady—flying horizontally at a uniform speed—these three forces must be in equilibrium. We can apply the parallelogram of forces. The wind force F must be exactly equal and opposite to the resultant of T and W (shown in the diagram by a dotted line).

The stability of an aeroplane.

The great problem in the design of an aeroplane is stability. Even on a calm day the air is never still, and little gusts and "air pockets" toss the plane about like a ship at sea.

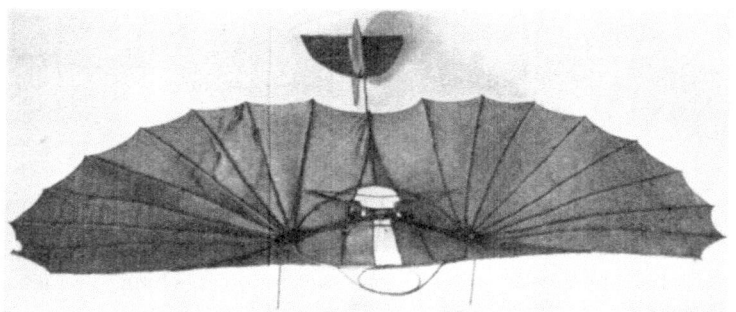

Science Museum, South Kensington, by permission of the R. Aeronautical Soc.
Fig. 160. Otto Lilienthal's glider.

Otto Lilienthal, a German, was the father of heavier-than-air flight. He made gliders, took them to the top of a hill, and ran into the wind until he could feel himself carried into the air. When caught by a gust he preserved his stability by throwing his body to one side of the glider to counteract the force. Unfortunately in 1896 he fell during one of his flights, and was killed.

The brothers Orville and Wilbur Wright, who were born in Dayton, U.S.A., continued the work. They made many experiments with gliders, and invented a method of preserving stability which has been used in all subsequent machines. They also fixed a motor cycle engine to one of their gliders, and on December 17th, 1903, they made the first successful flight in a power-driven aeroplane.

An aeroplane, like a submarine, needs to have a vertical rudder for sideways steering, and a horizontal rudder or elevator for up and down steering. Now when a cyclist rides round a corner at

Science Museum, South Kensington, by permission of Orville Wright, Esq.

Fig. 161. The Wright biplane, the first power-driven aeroplane
to fly successfully.

speed he needs to lean inwards or he will be upset; for the same reason the track at Brooklands is steeply banked. Thus when an aeroplane turns it will also be upset unless some force is applied to cause it to bank. The Wright brothers fixed to the back of the wings of their plane two hinged flaps, which could be moved up or down. These movable flaps are known as ailerons, and are used in all aeroplanes to-day. Look at the ailerons in the picture of the aeroplane in Fig. 162. When the aeroplane is steered to the left the right aileron is depressed, giving a bigger lift on that side,

Fig. 162. A large air-liner. Note the ailerons hinged to the back of the upper wing, and also the slots at the front of the wing, to prevent stalling at slow speed.

and the left aileron is raised, causing a smaller lift on the left.
In this way the aeroplane banks, and preserves its stability
when turning.

The sideways stability of the modern aeroplanes is increased
by making the wings slope slightly upwards from the centre of
the machine. Thus if the aeroplane tends to roll to one side, the
wing on that side becomes horizontal, and the air exerts on it a
greater lift while the lift on the other wing is decreased. Hence
the plane is automatically righted.

Structures.

Before a large steel structure, such as a bridge, is built, the
forces in each girder are carefully calculated in order that the
requisite strength may be determined. Before the Sidney Bridge
was built two years were spent in making calculations.

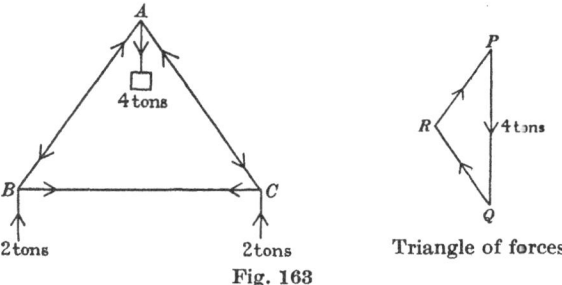

Fig. 163

The fundamental principle used is the triangle of forces.
Girders may either be in tension, that is to say, tending to be
stretched, when they are called *ties*, or in compression, when they
are called *struts*.

Calculation of forces in a simple structure. Consider a triangular
structure *ABC* which is vertical and supports at *A* a load of 4
tons. The forces in the members are represented by arrows in the
figure. *BA* and *CA* are pushing up at *A* to support the 4 ton
weight, and they are pressing downwards with an equal and
opposite force at *B* and *C*. The members *AB* and *AC* are pushing
outwards to prevent themselves from being compressed, and
they are therefore struts. On the other hand, *BC* is in tension,
and is a tie.

The vertical arrows at B and C represent the upward supporting force of the ground. They must each represent a force of 2 tons weight (if we ignore the weight of the framework), since the figure is symmetrical, and there is no reason why one of these forces should be larger than the other.

Consider the three forces acting at the point A. They are in equilibrium and hence can be represented in magnitude and direction by the sides of a triangle—the triangle of forces. All we need to do therefore to find the forces in BA and CA is to draw a triangle to scale.

Draw PQ parallel to the force of 4 tons wt. and of a proportionate length, say 4 in. Draw PR and QR parallel to BA and CA respectively to intersect in R. Now measure RP and QR, and their lengths will represent the forces in BA and CA.

Assuming ABC to be equilateral, draw the triangle of forces PQR accurately for yourself, and hence find the forces in AB and AC. [*Answer.* 2·31 tons wt.]

A triangle representing the forces at B or C should also be drawn, and hence the forces in BC determined. [*Answer.* 1·15 tons wt.]

Bridges.

Among the most interesting modern structures are steel bridges. They may be divided into four main types: (1) girder, (2) arch, (3) cantilever, (4) suspension, bridges.

Fig. 164. Simple plank bridge.

The simplest of all bridges is the plank. The main drawback is that it is apt to sag. It can be prevented from doing this by holding up the middle with a vertical rod attached to two inclined rods (see Fig. 165). Such a structure is known as a triangulated truss, and is similar to the structure we have just considered. It is light and strong, and owes its strength to its triangular shape, since a rigid triangle cannot be deformed without breaking or altering the length of one of its sides. Fig. 166 shows the Ava Bridge, Burma, one of the largest girder bridges in the world. Each span is formed of a number of triangles and is simply a development of the triangulated truss.

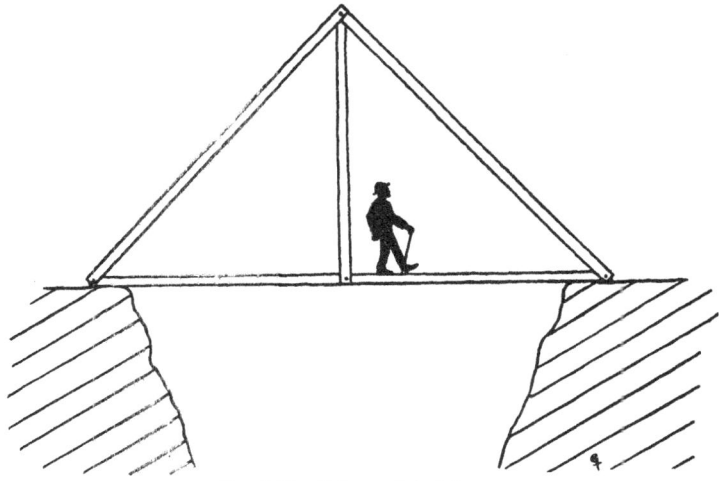

Fig. 165. Triangulated truss.

The arch is a stronger type of bridge than a girder bridge. Its thrust bears sideways on its abutments, i.e. it tends to push its supports outwards. The largest arch in the world is the Sidney Harbour Bridge, Australia, which has a main span of 1650 feet and was opened in 1932.

The finest bridge in the British Isles is a cantilever bridge— the Forth Bridge in Scotland. It has a total length of $1\frac{1}{2}$ miles. Two central spans are supported by overhanging arms called cantilevers. Fig. 169 is an illustration devised by Sir Benjamin Baker, one of the engineers of the Forth Bridge, to show the cantilever principle. The chief ties and struts are indicated in this diagram.

Fig. 166. The Ava Bridge in Burma. It was opened in 1934 after taking five years to build and is one of the largest girder bridges in the world.

Fig. 167. The Sydney Harbour Bridge, a magnificent example of an arch bridge. It has a main span of 1650 feet.

Fig. 168. The Forth Bridge, Scotland—a cantilever bridge.

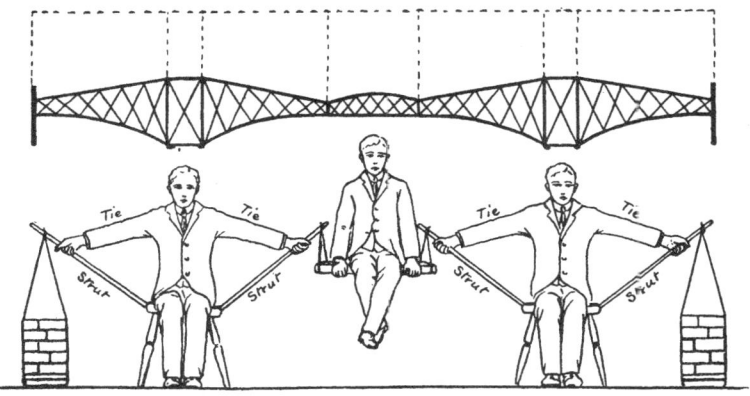

Fig. 169. Diagram illustrating the principle of the cantilever bridge.

The longest cantilever bridge in the world, the Quebec Bridge, took fifteen years to build as a result of two disasters. In 1907, five years after its commencement, the cantilever on the south bank was complete and about one hundred workmen were building out from it the central span. Suddenly the whole structure collapsed and fell into a heap of ruins. Some of the lower

By courtesy of Canadian National Railways

Fig. 170. Lifting the central span of the Quebec Bridge into position. This picture shows clearly how the two cantilever arms support the central span.

girders had been unable to bear the excessive strain imposed by the construction.

The design of the bridge was then modified. The central span was built separately, and after both cantilevers had been constructed, it was towed into the middle of the river on pontoons, ready to be lifted into place. It had been raised about 30 ft. when something broke and it fell with a crash into the water.

The engineers did not despair. They built a new span, and finally in 1917 the bridge was completed.

Another type of bridge is the suspension bridge, which consists of steel chains or cables hanging from one bank to the other. From these chains a level roadway is suspended. There are several fine examples of this type of bridge in Great Britain, notably at Bristol, spanning the Avon Gorge, and also between

Sport and General Press Agency

Fig. 171. The Brooklyn Suspension Bridge, New York. The roadway is supported from the hanging chains.

Wales and Anglesey, across the Menai Straits. The largest suspension bridges in the world are to be found in New York City— the Brooklyn, the Williamsburg and the Hudson River bridges. The last named has nearly twice the span of any other suspension bridge and its steel towers are 635 ft. high.

An interesting point in the design of the suspension bridge is the fact that the supporting chains or cables must be made to hang at the same angle on either side of the supporting towers. Thus the horizontal component of the pull of the chain on each side will balance the other; the tower will have to sustain a vertical force only and there will be no tendency for it to topple over.

In looking over the pictures and diagrams of bridges given in this book and elsewhere, try to decide which members are in tension and which are in compression, and where the weight of the structure is supported.

Summary

Velocity and force are vector quantities; they have direction as well as magnitude and can be represented therefore by straight lines. The following apply to all vectors: we shall state them for forces.

Parallelogram of Forces.

If two forces are represented in magnitude and direction by the adjacent sides of a parallelogram, then the diagonal (through their point of intersection) represents the resultant in magnitude and direction.

Triangle of Forces.

If three forces acting at a point are in equilibrium, they can be represented in magnitude and direction by the three sides of a triangle, taken in order.

Resolution of Forces.

A single force may be resolved into two components by drawing a right-angled triangle.

If three (non-parallel) forces are in equilibrium, they must intersect in a point.

QUESTIONS

1. (a) An aeroplane flies at 90 m.p.h. relative to the air, and heads due N., but it is carried out of its course by a wind blowing at 50 m.p.h. towards the N.W. Find its velocity relative to the ground (giving direction as well as magnitude).

(b) If the wind blows towards the S.W., what is the velocity of the aeroplane?

2. Two tug boats are pulling on a vessel, one with a force of 1200 lb. wt., the other with a force of 2000 lb. wt. The cables from the two tugs

are at right angles to one another. Find the resultant pull on the vessel.

3. A boy pulls a toboggan by means of a rope passing over his shoulder. His shoulder is 4 ft. from the ground, and the length of rope between his shoulder and the toboggan is 5 ft. Find the horizontal component of his pull if the tension of the rope is 40 lb. wt.

4. Explain how a kite flies. Draw diagrams in which the three forces acting on it are clearly marked, and discuss the conditions for them to be in equilibrium.

5. Explain why it is easier to pull a heavy roller than to push it.

6. An archer bends a bow until the two parts of the string make an angle of 90° with each other. If he pulls with a force of 120 lb. wt., find the tension in the string.

7. A body is moving N.E. at 10 ft. per sec. It is given a blow and as a result begins to move N.W. at the same *speed*. What *velocity* has been added to it by the blow to produce this change?

School Certificate Questions

8. State and explain the principle of the triangle of velocities.

A river 1 mile across has a current of velocity 4 m.p.h. In what direction must a launch, moving at 8 m.p.h., steer in order to land directly opposite the starting point? What will be the time for the crossing?

9. A motor cyclist is travelling due N. with a speed of 30 m.p.h. and a wind is blowing from the E. with a speed of 10 m.p.h. What is the apparent direction and velocity of the wind to the cyclist? Illustrate your answer with a diagram.

10. Explain the proposition known as the Parallelogram of Forces, and describe how it can be illustrated experimentally.

The ends of a rope 7 metres long are attached to two pegs A and B, 5 metres apart, the line AB being horizontal. A body weighing 500 gm. hangs from the rope at a point 3 metres from one end. What is the tension in the two parts of the rope?

11. Explain why it is so easy to break a tightly stretched string by pulling at it sideways.

A string stretched between gate posts, 6 ft. apart, snapped when its middle point was pulled out 2 ft. with a force of 15 lb. What was the greatest tension the string could stand?

12. State the law of equilibrium of three forces, the lines of action of which are not parallel.

An iron ball weighing 10 lb. hangs at the lower end of a long wire. What steady horizontal pull would be applied to the ball to keep the wire deflected 30° from the vertical? How would you test this experimentally?

13. Under what conditions is it possible for a body to remain at rest under the action of three forces in one plane? A ball weighing 6 oz. rests in a right-angled groove of which the two sides are equally inclined to the vertical. With what force does the ball press on one side of the groove?

14. Three forces of 1, 2, 3 lb. act on a particle in directions respectively parallel to the three sides of an equilateral triangle, taken in cyclic order. In what direction will the particle tend to be moved and what force will be required to keep it in place?

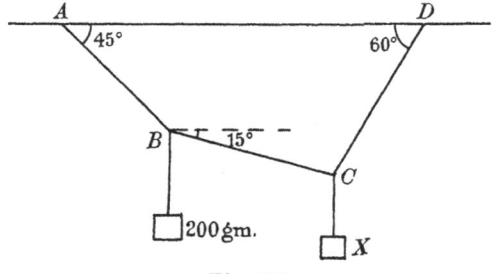

Fig. 172

15. What is meant by the resolution of a force? Show how a force may be resolved into two components parallel to any two required directions. Explain by means of diagrams how a sailing boat may be propelled by the wind in a direction about 45° with that from which the wind is coming.

16. Find the horizontal force required to keep a particle of mass 2 gm. at rest on a smooth plane inclined at 30° to the horizontal.

17. A smooth uniform sphere of mass 20 lb. and radius 6 in. is suspended by a string 1 ft. long, which is attached to a point on the surface of a smooth vertical wall. If the sphere rests against the wall, find the force exerted by the sphere on the wall.

18. A uniform rod, weighing 20 lb., hinged at its lower end, is held at an inclination of 30° to the vertical by a horizontal string attached

to its upper end. Find by construction or calculation the tension in the string and the reaction at the hinge.

19. What are the conditions of equilibrium of three forces acting in one plane?

A uniform rod AB of length 100 cm. and mass 200 gm. is hinged to a wall at A and is held horizontally by means of a string 200 cm. long connecting B with a point in the wall vertically above A. Find the tension in the string and the reaction at the hinge.

20. A string $ABCD$ is attached to a horizontal beam at A and D. A 200 gm. wt. hangs from B and an unknown wt. X from C. Find graphically (or otherwise) the value of X and the tensions in AB, BC and CD, the inclinations of the strings to the horizontal being as shown in Fig. 172.

Chapter X

GALILEO AND FALLING BODIES. MOTION

The swift advance of modern science dates from the sixteenth and seventeenth centuries, and the man who, above all others, helped to set the tide flowing was Galileo (1564–1642).

Galileo lived in the age of the Renaissance, a time of rebirth and quickening of the human spirit after a thousand years of slumber and stagnation. Spring, as it were, followed winter. Men were filled with a new vitality; they lived vividly, and made splendid progress in the Arts and Science.

One of the causes of the Renaissance was a revival of interest in the ancient Greek writers. As the Turks closed in upon Constantinople, the capital of the Byzantine empire, which they captured and sacked in 1453, the learned Greek scholars fled to Italy, taking their books and manuscripts with them.

Aristotle, the greatest of the ancient Greek writers, had compiled a comprehensive work containing all the knowledge of his time, and corresponding to our modern *Encyclopaedia Britannica*. But there was this great difference. The *Encyclopaedia Britannica* is revised and a new edition issued every ten or twenty years; Aristotle's encyclopaedia had not been revised for nearly 2000 years, for the simple reason that little new knowledge had been acquired. Its venerable age gave it tremendous authority. Scholars respectfully scanned its pages, annotated and wrote about it, but never dreamed of questioning the truth of any of its facts.

The young Galileo, at this time lecturer in mathematics in the University of Pisa, had the audacious daring to challenge one of its statements, namely that bodies fall to the ground with a speed which is proportional to their weight. This means that if bodies of 100 lb. and 1 lb. are pushed over a cliff the former falls 100 times as fast as the latter. Galileo decided to test the matter for himself by experiment. The results which he obtained have proved to be of profound importance in the study of motion.

When two objects, such as a piece of iron and a piece of cork, are dropped from the hands, they fall so quickly that it is very difficult to determine whether one reaches the ground before the other. Try this for yourself. It is therefore desirable that the weights be dropped from a considerable height. There is in Pisa a high leaning tower which is ideally suitable for the experiment.

Photo: Mansell

Fig. 173. The Leaning Tower of Pisa.

Galileo and an assistant dropped a cannon ball weighing over 100 lb. and a musket bullet weighing about $\frac{1}{2}$ lb. from the top, and found that they reached the ground practically simultaneously. By many such experiments he showed that, apart from discrepancies caused by air resistance, **all bodies fall to the ground at the same speed, whatever their weight.**

Aristotle was proved to be wrong; and if one of his facts was incorrect, others might be also. The Greeks, though one of the most intellectual races the world has seen, believed that the secrets of Nature could best be won by contemplation. They regarded experiment with disfavour, even with disdain. The barrenness of this attitude had to be proved before modern science was possible. It was Galileo's experiments at the leaning tower of Pisa which did this, and thereby initiated the era of scientific experiment.

Velocity and acceleration.

We must pause here in our account of the discoveries of Galileo to consider the meaning of several terms used in connection with motion.

Suppose a car is driven from Repton to London, a distance of about 120 miles, in 5 hours. The average speed for the journey is $\frac{120}{5} = 24$ m.p.h.

$$Average\ speed = \frac{Distance}{Time}.$$

The speed at different points on the journey, however, will vary. In traffic the speed may fall below 10 m.p.h. and on the open road it may exceed 50 m.p.h. The *speed at any instant* (for the full meaning of this term, see p. 229) may be determined approximately by reading the speedometer.

If the car moves with a steady unchanged speed, say 40 m.p.h., for a period, it is said to be travelling with *uniform speed*. The speed at every instant during this period must be the same. The seconds finger of a watch, therefore, does not move with uniform speed, since its motion is jerky.

Another term, velocity, is used to denote speed in a given direction. *Velocity is the distance traversed in unit time in a given direction.*

As we saw in Chapter IX, velocity is a vector quantity, and can only be fully represented by a straight line. Thus while we may speak of a body moving with uniform speed in a circle, it would be incorrect to say that the body moved with a uniform velocity, since it is constantly changing in direction. A body moving with uniform velocity must be travelling in a straight line.

When a car starts from rest its velocity is increasing and it is said to be accelerating.

$$Acceleration = \frac{Increase\ in\ velocity}{Time}.$$

Thus if a car attains a velocity of 40 m.p.h. from rest in 2 min., its average acceleration is $\frac{40}{2} = 20$ m.p.h. per minute.

A car on a journey is constantly changing its velocity and therefore constantly accelerating and "decelerating" or retarding. A powerful car can get up speed rapidly, i.e. it has high acceleration.

Example. A small sports car can attain a velocity of 60 m.p.h. from rest in 38 sec. Find its average acceleration in ft. per sec. per sec.

$$60 \text{ m.p.h.} \quad = \frac{60 \times 1760 \times 3}{60 \times 60} \text{ ft. per sec.}$$

$$= 88 \text{ ft. per sec.}$$

$$\therefore \text{ Acceleration} = \tfrac{88}{38} \text{ ft. per sec. per sec.}$$

$$= 2 \cdot 32 \text{ ft. per sec. per sec.}$$

If a body gains velocity at a steady rate, it is said to have a *uniform acceleration*.

Example. A body starts from rest with a uniform acceleration of 2 ft. per sec. per sec. What will be its velocity after 30 sec. and how far will it have gone?

Increase in velocity per sec. $= 2$ ft. per sec.

Velocity after 30 sec. $\qquad = 2 \times 30$

$\qquad\qquad\qquad\qquad\qquad = 60$ ft. per sec.

Average velocity $\qquad\qquad = \tfrac{1}{2}$ (velocity at beginning + velocity at end)

$\qquad\qquad\qquad\qquad\qquad = \tfrac{1}{2}\ (0 + 60)$

$\qquad\qquad\qquad\qquad\qquad = 30$ ft. per sec.

$\therefore$ Distance gone $\qquad\qquad = $ Average velocity $\times$ time

$\qquad\qquad\qquad\qquad\qquad = 30 \times 30$ ft.

$\qquad\qquad\qquad\qquad\qquad = 900$ ft.

The law governing the fall of bodies.

Galileo had demonstrated that all bodies fall at the same speed, whatever their weight. He now asked whether, since all bodies fall in the same way, there was any simple law governing their fall. His greatness lay, not so much in his solution of this question, but in the fact that he asked it. For science was then in its infancy, and it was not realised that all the varied phenomena of Nature could be accounted for by a series of simple laws.

Galileo began to speculate on possible laws that might hold. It was obvious that the distance traversed by a falling body was not proportional to the time, for the fall of a body becomes faster and faster, and it covers longer and longer distances in unit time. Since, however, a body gathers speed as it falls, Galileo postulated that the velocity might increase regularly with the time, in other words, that a falling body might have a uniform acceleration. The problem was how this could be proved.

Bodies took only two or three seconds to fall, even from the top of the leaning tower of Pisa, and as watches had not been invented, Galileo found it impossible to make accurate measurements of the times taken to fall different distances vertically. His only timepiece was a water clock, which consisted of a large tank with a small hole in the bottom over which he put his finger. Intervals of time were measured by the volume of water run out.

The difficulty was met by allowing a ball to roll quite slowly down an incline. Galileo reasoned that the distances rolled down the incline in different times would all be proportional to the distances the ball would fall vertically in those times.

He thus obtained a series of pairs of values of distances and times. He had then to find whether these facts fitted in with his theory. He did it in the following way:

Suppose the velocity of a body increases by the same amount in every second, i.e. it has a uniform acceleration, and that it starts from rest.

Velocity of body at beginning $= 0$.

Let Acceleration $= a$ cm. per sec. per sec.

Then Velocity after t sec. $= at$ cm. per sec.

$\therefore$ Average velocity during this time $= \frac{1}{2}$ (velocity at beginning
$\qquad\qquad\qquad\qquad\qquad\qquad\qquad\qquad + $ velocity at end)

$$= \frac{0 + at}{2} \text{ cm. per sec.}$$

$$= \tfrac{1}{2}at \text{ cm. per sec.}$$

Distance travelled $=$ Average velocity
$\qquad\qquad\qquad\qquad\qquad\qquad\qquad\qquad \times$ time.

Let   Distance travelled $= s$ cm.

$$\therefore \quad s = \tfrac{1}{2}at \times t \text{ cm.}$$

$$s = \tfrac{1}{2}at^2 \text{ cm.}$$

By the above reasoning Galileo showed that if his theory were correct, if, that is to say, a body starting from rest obtains equal increases of velocity in equal times, then the distance it goes will be proportional to the square of the time that it takes.

Galileo's theory was thus expressed in terms of two measurable quantities, distance fallen and time taken. He proceeded to find whether his theory was correct by examining his figures. He found that his readings of the distance were proportional to the square of the times taken, and so his theory was proved. [This can be done most simply by plotting a graph of distance against (time)^2 which is found to be a straight line passing through the origin.]

Hence Galileo proved by his experiment that **a falling body has a uniform acceleration.** This is known as **the acceleration due to gravity.** It is denoted by g.

Since Galileo's time the acceleration of a vertically falling body has been found accurately.

$$g = 32 \text{ ft. per sec. per sec.}$$
$$= 981 \text{ cm. per sec. per sec.}$$

Fig. 174 is a photograph of the fall of a body. Snapshots lasting about $\frac{1}{100}$ sec. were taken, at intervals of $\frac{1}{30}$ sec., of a brightly illuminated falling ball. Notice that in each successive $\frac{1}{30}$ sec. the ball falls a greater distance than in the previous $\frac{1}{30}$ sec. After falling, the ball bounced first on a tilted steel plate and then on a wooden board. It will be seen that the ball loses speed when rising at the same rate as it gains speed when it falls. The curved trajectory described by the ball after bouncing is called a parabola.

Calculation of the distance gone each second by a falling body.
Suppose the body falls from rest.

1.   Velocity at beginning $= 0.$

                 Velocity at end of 1st sec. $= 32$ ft. per sec.

     ∴   Average velocity $= \dfrac{32 + 0}{2}$

                                              $= 16$ ft. per sec.

Fig. 174. A photograph of a falling and bouncing ball taken by Mr D. G. A. Dyson. A small steel ball was coated with magnesium oxide smoke, very brightly illuminated against a black background, and held and released by an electromagnet. Exposures of $\frac{1}{100}$ sec. were made every $\frac{1}{30}$ sec. by rotating in front of the camera lens a vertical disc having holes spaced evenly round its circumference and turned by a gramophone motor. The ball bounced first on a tilted steel plate and then on wood. The faint outline of light inside the larger parabola is the rim of a cardboard shield placed to hide the lamps with which the ball was illuminated.

∴ Distance travelled   = Average velocity × time

$$= 16 \times 1$$

$$= 16 \text{ ft.}$$

2.   Velocity at end of 2nd sec.   = 64 ft. per sec.

∴ Average velocity $= \dfrac{64 + 0}{2}$

$$= 32 \text{ ft. per sec.}$$

∴ Distance travelled $= 32 \times 2$

$$= 64 \text{ ft.}$$

∴ Distance travelled in 2nd sec. = Distance in 2 sec.

                            − distance in 1 sec.

$$= 64 - 16$$

$$= 48 \text{ ft.}$$

Study the table given below and check, on a piece of paper by the methods shown above, all the figures given.

Time (sec.)	Velocity at end of time (ft. per sec.)	Distance gone in whole time (ft.)	Distance gone in each sec. (ft.)
0	0	0	0
1	32	16	16
2	64	64	48
3	96	144	80
4	128	256	112
5	160	400	144

Note especially that the velocity increases by 32 ft. per sec. each second.

Example 1. *A boy throws up a ball and it remains in the air for* 3 *sec. How high did it rise?* A body takes the same time to rise a certain height as to fall the same distance. Thus the ball is rising for $1\frac{1}{2}$ sec. and falling for $1\frac{1}{2}$ sec.

Hence we only need to find how far a body falls in $1\frac{1}{2}$ sec.

Velocity at top   = 0.

Velocity at bottom $= 32 \times 1\frac{1}{2}$

$$= 48 \text{ ft. per sec.}$$

$\therefore$ Average velocity $= \dfrac{0+48}{2}$

$= 24$ ft. per sec.

$\therefore$ Distance fallen $=$ Average velocity $\times$ time

$= 24 \times 1\frac{1}{2}$

$= 36$ ft.

Example 2. *A man makes a* 100 *ft. dive from scaffolding on a pier. How long does he take to reach the water and what is his velocity as he enters it?* We have shown that

$s = \frac{1}{2}at^2,$

where $\quad s =$ Distance travelled $= 100$ ft.,

$a =$ Acceleration $\quad = 32$ ft. per sec. per sec.,

$t =$ Time (in sec.).

$\therefore \quad 100 = \frac{1}{2} . 32 . t^2,$

$t = \sqrt{\frac{100}{16}} = \frac{10}{4} \qquad = 2\frac{1}{2}$ sec.

$\therefore$ Velocity $=$ Acceleration $\times$ time

$= 32 \times 2\frac{1}{2}$ ft. per sec.

$= 80$ ft. per sec.

Calculate the value of this velocity in m.p.h.

Air resistance.

Everyone knows that if a penny and a piece of paper are dropped from the same height, the penny will reach the ground much more quickly than the paper. The air resistance on the paper is nearly as great as its weight, whereas in the case of the penny it is almost negligible. Now if a small piece of paper is cut out, rather less than the size of the penny, and allowed to rest on the penny, when they are dropped together with the penny horizontal they will fall at the same speed. The penny overcomes the air resistance for both itself and the paper. This is an experiment worth trying.

After the air pump had been invented, Sir Isaac Newton performed an experiment with a guinea and a feather. He allowed them to fall *in vacuo* and found that they fell with exactly the same speed. The experiment may be repeated with small pieces of paper and brass in a long wide glass tube which has been evacuated.

Galileo himself recognised the effect of air resistance. He investigated the matter by allowing bodies to fall through water, in which the effect is exaggerated.

Terminal velocity.

A body falling in air continues to accelerate until it reaches a certain velocity, known as its *terminal velocity*, and then it falls steadily, without any further increase in speed. The faster a body falls in air, the greater the air resistance, and at the terminal velocity the air resistance has reached a value exactly equal to the weight of the body.

The terminal velocity of snow is only a few ft. per sec., and, fortunately for us, that of rain and hail is not sufficiently high to cause serious damage. Insects can fall without harm from very great heights, and now, with the aid of a parachute, so also can man.

The terminal velocity of a man falling through air without a parachute is about 120 m.p.h.: with a parachute, his terminal velocity is equal to that acquired by jumping from a 13 ft. wall, i.e. about 20 m.p.h.

Why bodies fall.

Aristotle said that bodies fall because they are seeking their natural place. He considered it the natural thing for bodies to fall downward. But what is downward? Now that we know the earth is round and not flat we realise that cricket balls falling in England and Australia are moving in almost opposite directions. The people, too, in Australia walk with their heads pointing in, what is to us, a downward direction.

How are we to account for this fly-like ability of the Australians, and the fact that they are so seldom troubled with a rush of blood to the head?

The reason is that there is no fixed downward direction in space. Bodies on or near the earth always fall towards its centre, and this is the downward direction at each place. They fall because the earth attracts them; the attraction is called *the force of gravity*.

Let us imagine a lift shaft from the British Isles through the centre of the earth to the Antipodes, or (with a little bending) to Australia. If the lift falls freely, it will acquire a tremendous

Fig. 175. With the aid of a parachute a man can fall at a terminal velocity of about 20 m.p.h.

velocity, overshoot the centre of the earth, and rise up on the other side. As it passes the earth's centre, the "downward" direction is suddenly reversed. The ceiling becomes the floor, and if the lift is braked and brought to rest, the passengers will fall head-first on to the ceiling.

The moon is much smaller than the earth, and its attraction for bodies is only one-sixth that of the earth. Bodies on the moon therefore fall only one-sixth as quickly as on the earth. The high jump record on the moon would be over 30 ft., and the cricket ball could be thrown more than a quarter of a mile. On the other hand, the planet Jupiter is considerably larger than the earth, and attracts bodies with a force 2·65 times as great as the earth. The high jump record on Jupiter would therefore be about 2 ft., and ordinary walking would be laborious and exhausting. A fall from a high wall would probably be fatal.

Locating oils and metal ores by variation of g. The acceleration due to gravity, *g*, varies slightly at different places on the earth's surface owing to the earth not being a perfect sphere. However, there are also slight local variations due to the nature of the rocks beneath the surface. By measuring these variations it is possible to detect the presence of oil and heavy metal ores. This has an important practical application and is known as geophysical prospecting.

The life of Galileo.

As might be expected of a great man living during the stirring times of the Renaissance in Italy, Galileo's life was dramatic, but its end was one of the most moving and tragic in the history of science.

In 1592 he left Pisa, where he had performed his experiments on falling bodies, and became a professor at Padua, a post which he held for eighteen years. During this time he invented a telescope and with its aid made many important astronomical discoveries. He saw for the first time that the planet Jupiter has four moons which revolve round it, and that there are spots on the sun which appear and disappear at intervals, suggesting that the sun is rotating. He became an ardent supporter of the theory that the earth is revolving round the sun. This theory had been revived by the monk Copernicus some fifty years earlier, and had been denounced by the Roman Catholic Church as heresy.

Fig. 176. A man who can jump to a height of 5 ft. on the earth would be able to clear 30 ft. on the moon but no more than 2 ft. on the planet Jupiter.

It is interesting to consider the Church's case. All through the Dark Ages from the fall of the Roman Empire to the Renaissance, the Christian Church had stood between mankind and a relapse into barbarism. For more than a thousand years her authority in the intellectual as well as the moral sphere had remained unchallenged. During this period, certain beliefs, such as the earth as the centre of the universe, and the geographical position of

Fig. 177. Galileo (from a portrait in the *Bodleian Library*).

heaven and hell, had become an established part of the Church's dogma. It was regarded as the height of pride and presumption for one man to set his opinion against the ancient teaching of the Church; it was believed that such action would lead inevitably to his own damnation and mortal sin in others.

Galileo was warned repeatedly against teaching the Copernican theory, but it was not until he published a most powerful and widely read book in support of it that the Church took action. He was then, at the age of seventy, summoned to Rome, and tried by the Inquisition. None of his friends dared support him, and for months his trial dragged on. The formal stages of examination by the Inquisition are as follows:

1st. The official threat in the Court.

2nd. The taking to the door of the torture chamber and renewing the official threat.

3rd. The taking inside and showing the instruments.

4th. Undressing and binding upon the rack.

5th. Territio realis.

"Through how many of these ghastly acts Galileo passed I do not know. I hope and believe not the last.

"There are those who lament that he did not hold out and accept the crown of martyrdom thus offered to him. Had he done so, we know his fate—a few years' languishing in the dungeons, and then the flames." (*Pioneers of Science*, Sir O. Lodge.)

"An utterly broken and disgraced old man" Galileo signed a document recanting all his teaching and denying his discoveries. This document was sent throughout the length and breadth of Europe, and read publicly in the churches and universities.

There is a legend that, at his recantation, Galileo muttered, under his breath "E pur si muove" (It does move all the same).

Soon afterwards he became blind and his last years were spent quietly in seclusion.

Distance-time graph.

In this chapter we have dealt solely with falling bodies, and these have a uniform acceleration. But motion in general, that for instance of a runner, motor car or train, is not so simple, and it may conveniently be studied by means of graphs.

Suppose a mile runner carries a stop-watch, and notes the time every 200 yd., or, better still, that he has observers posted to record his times. He can then plot a distance-time graph. Fig. 178 represents a typical graph. It will be seen that the slope of the graph varies. This indicates that the speed of the runner is not uniform. The graph flattens out in the middle, showing that his speed drops, and then gets steeper again at the end, showing a spurt in the last lap. If the runner had kept up a uniform speed throughout (and this is the most economical way of running), the graph would have been a straight line. *In order to find his speed or velocity at any instant, the slope of the graph at that point must be found by drawing a tangent.* Thus

$$\text{Velocity at } P = \frac{AB}{PB}.$$

We can see that this must be so from the following argument. If the runner goes from P to Q (on the graph), he has traversed a distance BQ in a time PB. Hence his average velocity is QB/PB.

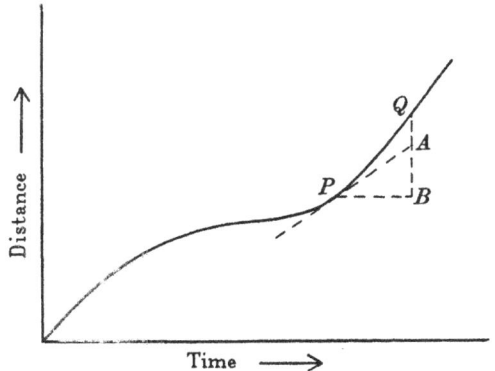

Fig. 178.   Distance-time curve.

But the velocity is changing all the time from P to Q, and to find the velocity at P we must take Q extremely close to P, when the chord PQ will coincide in direction with the tangent PA. Hence

$$\text{Velocity at } P = \frac{AB}{PB}.$$

At Cornell University, U.S.A., a method of electrical timing is used so that the times of a runner at intervals of a few yards at first, and then every 10 yards in a 100 yards race, are automatically recorded. Alongside the running track, at measured intervals, are placed coils of insulated copper wire, which are connected by wires to an instrument for detecting electric current, called a galvanometer, in the pavilion. The runner carries a magnetised strip of steel round his body and as he passes each coil the magnet causes a current to flow in the coil. This current

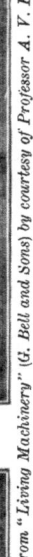

(a)

(b)

From "Living Machinery" (G. Bell and Sons) by courtesy of Professor A. V. Hill

Fig. 179 (a). The running track at Cornell University, U.S.A. Electric currents are induced in the coils at the side of the track as the runner passes them, thus enabling him to be timed very accurately at different points on the track.

Fig. 179 (b). Runner with magnetised steel belt which induces the electric currents in the coils.

affects the galvanometer practically instantaneously, and a record is made photographically over a steadily moving strip of bromide paper.

Velocity-time graph.

It will be noted that a distance-time graph gives no information about acceleration. A velocity-time graph is necessary for this purpose. Suppose a man sits beside the driver of a racing car and makes a note of the reading of the speedometer at con-

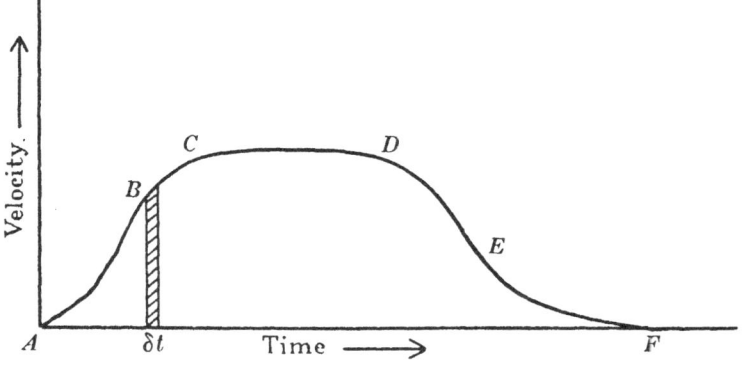

Fig. 180. Velocity-time curve.

venient intervals of time from start to stop. He can then use his readings to plot a velocity-time graph. *The slope of the graph at any point, obtained by drawing a tangent, gives the acceleration at that instant.* In Fig. 180 it will be seen that the slope of the graph, and therefore the acceleration of the car, at first increases from A to B, then decreases from B to C, and is zero from C to D (when the velocity is constant). The portion DEF represents the motion while the car is being braked, during which time the car has a negative acceleration, or retardation. The retardation increases from D to E and then decreases from E to F.

It is possible to find the distance gone by the car from the velocity-time graph. If the velocity had been constant, the distance could have been obtained by multiplying the velocity by the time. But the velocity varied continually. Consider therefore a very short interval of time (marked δt on the graph) during which the velocity did not change appreciably. The distance gone in this short time is the product of the velocity and the interval of time, which is equal to the area of the shaded strip. We can divide the whole time into a large number of such short intervals and the total distance gone will be equal to the sum of the areas of strips which will cover the whole area under the graph. *Thus the distance gone is represented by the area between the graph and the time axis.*

Equations of motion for a uniformly accelerated body.

The velocity-time graph for a body moving with a uniform acceleration (such as a falling body) will be a straight line (see

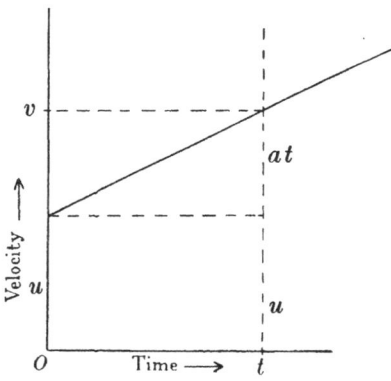

Fig. 181. Velocity-time curve.

Fig. 181). If the body does not start from rest, the graph will not pass through the origin.

Let $u =$ Initial velocity of body,

$\quad\quad v =$ Velocity after time t,

$\quad\quad a =$ Acceleration.

Then $v = u + at.$ ......(1)

Distance gone in time t

       = Area under graph

       = Area of rectangle + area of triangle,

$\therefore\ s = ut + \tfrac{1}{2}at^2.$ ......(2)

We can deduce from (1) and (2) a third useful equation which does not contain t.

From (1)

$$t = \frac{v-u}{a}.$$

Substituting for t in (2),

$$s = u\,\frac{v-u}{a} + \tfrac{1}{2}a\left(\frac{v-u}{a}\right)^2.$$

$$\therefore\ 2as = 2uv - 2u^2 + v^2 - 2uv + u^2,$$
$$v^2 = u^2 + 2as. \qquad\qquad\qquad\dots\dots(3)$$

Experiment. To show that a ball or flywheel rolling down an inclined plane has a uniform acceleration, i.e. a steady increase in velocity. Allow a ball to roll down an inclined groove, or a flywheel to roll down rails. A flywheel is better because it rolls more slowly. Obtain a series of values, about six, of distance rolled and time taken.

Fig. 182. Flywheel rolling down an inclined plane.

Proceed as follows. Start the flywheel by removing a ruler suddenly from in front of it. Place a block of wood across its path and find the time taken to reach the block by means of a stopwatch, or, alternatively, if the timing is done with a metronome, move the block so that the wheel hits it simultaneously with a tick of the metronome. Measure the distance from the block to the starting point. Obtain at least three values of each pair of readings of distance and time, and take the average.

The object of the experiment is to prove that the velocity of the flywheel increases steadily. The method of proof, therefore, will have to be indirect. Now it has been shown (p. 219) that if a body moves with a steadily increasing velocity, i.e. a constant acceleration,

$$s = \tfrac{1}{2}at^2,$$

where
$s = $ Distance travelled,
$t = $ Time,
$a = $ Acceleration.

We need to find out therefore whether the distance gone by the ball or flywheel is proportional to the (time)².

Plot a graph of distance against (time)². The points will be observed to lie very nearly on a straight line. Draw a straight line through the middle of the points so that they are distributed evenly on either side of it. What does this prove?

Draw also a graph of distance against time. How and why does this differ from the other graph you have drawn?

SUMMARY

Speed = Distance gone in unit time.
Velocity = Speed in a given direction.
Acceleration = Gain in velocity in unit time.

A falling body has a constant acceleration, known as the acceleration due to gravity g, which is independent of its weight, so long as air resistance is negligible.

$$g = 32 \text{ ft. per sec. per sec.}$$
$$= 981 \text{ cm. per sec. per sec.}$$

Bodies fall towards the centre of the earth because the earth attracts them.

Equations of uniformly accelerated motion.

$$v = u + at,$$
$$s = ut + \tfrac{1}{2}at^2,$$
$$v^2 = u^2 + 2as.$$

Distance-time and velocity-time graphs.

Motion which is not uniformly accelerated may be conveniently studied by means of graphs.

Speed at any instant

$$= \frac{\text{Distance traversed in a very short interval of time}}{\text{Time interval}}.$$

It may be found by drawing a tangent to the distance-time graph and finding its slope (at that instant). The slope of the tangent to the velocity-time graph gives acceleration. The area between the velocity-time graph and the time axis gives the distance gone.

QUESTIONS

1. Define velocity. Convert a velocity of 60 m.p.h. into ft. per sec. [The answer to this question is worth memorising. It will enable you to simplify other similar conversions.]

2. Define acceleration.

A train, accelerating steadily, reaches a speed of 60 m.p.h. from rest in 20 min. Find its acceleration in (a) m.p.h. per min., (b) m.p.h. per sec., (c) ft. per sec. per sec.

3. A ball rolls down an incline and in the first sec., having started from rest, it goes 6 cm.

(a) What is its average velocity during the first sec.?

(b) What is its average velocity at the end of the first sec.?

(c) What is its acceleration?

(d) What will be its velocity at the end of the next sec.?

(e) How far will it have gone in 2 sec.?

(f) How far will it have gone during the second sec.?

(g) What will be its velocity and how far will it have gone at the end of 5 sec.?

4. A stone is dropped down a well, and the splash is heard after 4 sec.

Assuming the sound to have travelled instantaneously, find the depth of the well.

5. A boy dives from a height of 9 ft. How long does he take in falling, and what is his velocity on reaching the water?

6. A juggler throws up a ball to a height of 4 ft. with one hand, and catches it in the other. How long is the ball in the air?

7. With what vertical velocity must a cricket ball be thrown up to rise 100 ft.?

8. Criticise the following: "We reflected on looking over the fearful precipices that, one false step, and we should have fallen 10,000 ft. in 10 sec." Calculate (*a*) the time taken to fall 10,000 ft., (*b*) the distance fallen in 10 sec., (*c*) the necessary value of the acceleration due to gravity for a body to fall 10,000 ft. in 10 sec.

9. A stone is thrown down from the top of a cliff with a velocity of 20 ft. per sec., and reaches the bottom in 3 sec. How high is the cliff?

10. An aeroplane travelling at 90 m.p.h. drops a bomb. Assuming that the bomb continues to move forward horizontally at 90 m.p.h. while it is falling, plot a graph of its fall. [Calculate the vertical distance fallen, and the horizontal distance moved, in ft. at the end of each sec. up to 5 sec., and plot vertical distance against horizontal distance.]

11. Can a body at any moment have (*a*) uniform speed but changing velocity, (*b*) uniform velocity and steady acceleration, (*c*) uniform acceleration but no velocity?

Give reasons in each case. If your answer is yes, give examples to illustrate it.

School Certificate Questions

12. A tape measure is hung vertically and a bullet dropped from the zero mark. A snapshot of the falling bullet is taken and shows that while the shutter is open the bullet falls from the 16 ft. mark to 16 ft. 4 in. How long is the shutter open?

13. A juggler is maintaining four balls in motion, making each in turn rise to a height of 9 ft. from his hand. With what velocity does he project them and where will the other three balls be at the instant when one is just leaving his hand?

14. What do you understand by acceleration? Draw and explain velocity-time graphs to illustrate the difference between uniform and non-uniform acceleration.

A car starts from rest and maintains a uniform acceleration of 2 ft. per sec. per sec. in a straight track. As it starts, another car, moving with a uniform velocity of 30 m.p.h., passes it in the same direction on a parallel track. When will the cars be opposite each other again? What will then be the velocity of the first car?

15. Describe the path of a body projected horizontally under gravity.

A projectile dropped from an aeroplane moving horizontally with a velocity of $87\frac{3}{11}$ m.p.h. reaches the ground in 4 sec. Find (*a*) the horizontal distance of the point of impact from the point of release, (b) the height of the aeroplane, and (c) the velocity of the bomb when it strikes the ground.

16. A stone is dropped from the top of a cliff 400 ft. high. Draw a displacement-time curve for the stone, and use it to deduce the velocity of the stone at the instant when it is 150 ft. above the ground.

17. A body is projected vertically upwards with a velocity of 100 ft. per sec. Plot a velocity-time diagram for the first 4 sec. of the motion and from it deduce (a) the time at which the body reaches its highest point, (b) the greatest height attained.

18. The speedometer readings of a motor car, taken at half-minute intervals, are recorded in the following table:

Time (min.)	0	$\frac{1}{2}$	1	$1\frac{1}{2}$	2	$2\frac{1}{2}$	3	$3\frac{1}{2}$
Speed (m.p.h.)	5	5	5	10	15	$17\frac{1}{2}$	20	$22\frac{1}{2}$

Time (min.)	4	$4\frac{1}{2}$	5	$5\frac{1}{2}$	6	$6\frac{1}{2}$	7
Speed (m.p.h.)	25	$27\frac{1}{2}$	30	30	30	15	0

Plot the velocity-time graph for the car. From the graph determine (a) the acceleration at time 4 min., (b) the total distance travelled in 7 min. Explain clearly in each case how the value is obtained from the graph.

19. A body starts from rest and moves for 10 sec. with a uniform acceleration of 5 cm. per sec.^{2}; for the next 20 sec. it moves uniformly with the velocity acquired and it is finally brought to rest with a uniform retardation of 5 cm. per sec.^{2} Draw the complete velocity-time graph for the body, and *from the graph* determine (a) the total distance travelled, and (b) the time taken.

Chapter XI

THE LAWS OF MOTION. SIR ISAAC NEWTON. GRAVITATION

In the preceding chapter we considered falling bodies and explained the cause of their motion as the pull or attraction exerted upon them by the earth.

But there are other forces besides the force of the earth's pull, and motion is not always towards the earth's centre.

Sir Isaac Newton studied more generally the effect of forces in producing motion, and summarised the facts which he discovered in the form of three statements, known as Newton's Laws of Motion.

First Law of Motion.

A body continues in a state of rest or to move with a steady velocity in a straight line if it is not acted upon by forces.

Everyone knows that bodies do not suddenly start moving unless a force acts upon them; but that they continue to move with undiminished speed when the force ceases to act seems contrary to experience. A vehicle, once its motive power is cut off, always slows down and comes to rest. However, when the engine is disconnected and a car is allowed to freewheel, all the forces acting upon the car have not been eliminated. There are forces such as friction and air resistance opposing its motion.

Now the more the force of friction is reduced, the more slowly does a body lose speed. A stone sent skimming over a sheet of ice travels much farther than it does on a road; the friction between the stone and the ice is less than between the stone and the road. If the ice is swept to make it smoother, the stone will go farther still. It is reasonable to suppose that could the ice be made perfectly smooth, and friction entirely eliminated, the stone would have no tendency to slow down, but would continue in a straight line with a steady velocity.

It is a fact of common experience that bodies do not change their direction of motion unless acted upon by a force. When a car turns a corner the force to change its direction of motion is provided by the friction between the wheels and the ground. If a large enough force cannot be exerted on the tyres by a slippery road, the car skids owing to its tendency to continue to move in a straight line.

A stone which is being whirled round in a circle on the end of a string describes its curved path because the pull of the string is causing it continually to change direction. If the string is cut, the stone flies off at a tangent in what we believe would be a straight line if there were no force of gravity.

The fact is that we have no experience of bodies which are not under the action of forces. On or near the earth gravity cannot be eliminated, nor can friction.

Now the earth and the planets are moving round the sun: they move in a curved path because the sun is attracting them. No retarding force such as friction acts upon them, however, and the fact that no noticeable slowing up has taken place for thousands of years may be regarded as strong evidence in favour of Newton's first law.

The law is essentially a "limiting" law. The nearer we approach to the case of a body moving under the action of no retarding force—by reducing friction for instance—the more nearly is the law obeyed.

Mass.

The tendency of a body to remain at rest or to move with unchanged velocity is due to what is called its *mass* or *inertia*. If a body is massive, i.e. has a large mass, we find it hard to change its motion. Thus a cannon ball is much more difficult than a ping-pong ball to start moving or to stop; it has a greater mass.

Again, to find which of two barrels is full and which is empty, we need neither open nor lift them. A push with the foot will decide, for the full barrel, having the greater mass, needs a greater force to set it moving.

The mass of a body, then, *is its tendency to resist a change of motion*. It is measured in lb. or gm., the same units as weight. The masses of bodies are compared with the mass of the British Standard Pound or of the International Kilogram. Thus a body of mass 2 lb. is twice as difficult to start moving or to stop as the British Standard Pound.

The mass and weight of a body, however, are entirely different properties. To avoid confusion we call the unit of mass the lb. and the unit of weight (or force) the lb. wt. It is possible to measure them in the same units because *at any particular place the masses of all bodies are proportional to their weights* (i.e. the pull of the earth upon them). We shall explain why this is so later.

A heavy flywheel exhibits clearly the effect of its mass when its weight has been completely counteracted. The weight of the flywheel is supported by the bearings, and friction can be reduced to a comparatively small force by means of ball bearings. But the wheel offers a considerable resistance to being started or stopped revolving.

The weight of a body, i.e. the pull or attraction exerted by the earth upon it, changes with its distance from the centre of the earth. But its mass never varies; it is always the same. A body weighs more at the poles than at the equator, since at the poles it is nearer the centre of the earth. If it could be taken millions of miles away from the earth its weight would become very small indeed, but its mass would be unaltered. It would be just as hard to start moving or to stop.

Thus mass is a fundamental and inseparable property of matter; weight is a secondary and variable property, like temperature.

The short story by Mr H. G. Wells, *The Truth about Pyecraft,* while impossible and fantastic, draws an interesting distinction between mass and weight. Mr Pyecraft, "the fattest clubman in London", bemoans his enormous mass but always speaks of reducing his weight. Unfortunately an ancient Hindu remedy fulfils his spoken desire too literally. He loses his weight while his mass is unaffected, and he now has to weight his clothing with lead to prevent himself floating up into the air.

It is interesting to note that an ordinary balance measures mass, or, more accurately, compares the weights of two masses, the standard and that of the body being weighed; it is incapable of detecting a body's change of weight at different places on the earth's surface, since the standard "weights" are similarly affected. A spring balance, on the other hand, measures weight, since its extension indicates the pull of the earth for a body.

Instances of the effect of mass, or the tendency of a body to resist having its motion changed, are numerous, and may sometimes be spectacular. If your bicycle is stopped suddenly by a brick, your body tends to continue moving over the handle bars until brought to rest by the action of the earth's force upon you. If you step off a moving vehicle, your feet are stopped by the ground, but the rest of your body tends to continue moving, and you are liable to pitch forward. In order to keep upright you should move your feet quickly at first, and bring yourself to rest gradually. An inspector, when alighting from a bus, always leans backwards, so that the forward motion of his centre of gravity tends to bring him up to the vertical position again. If you are standing in a tram which starts suddenly, you are apt to fall backwards. A train rounding a curve tends to go straight on and would be derailed were it not for the flanges on its wheels and the banking of the track. Cyclists and motorists who attempt to take corners too quickly topple outwards.

The effect of forces in producing motion. A body requires no pushing in order to keep it moving uniformly but only to make it go faster. The first law of motion tells us that when no force acts on a body its velocity is unchanged and the second law of motion tells us how the velocity of a body changes when a force does act upon it.

A constant steady force causes a body's velocity to increase (or decrease) steadily, in other words, to produce a uniform acceleration.

Second Law of Motion.

When a force acts on a body it produces an acceleration which is proportional to the magnitude of the force. [See also p. 255.]

When a body is falling freely, its weight is causing its mass to move with a constant acceleration g.

We can quite simply calculate the acceleration produced in a body by any force. Suppose a force of F lb. wt. acts on a body of mass W lb.

Now, W lb. wt. produces in a mass of W lb. an acc. of g ft./sec.^{2}

$$\therefore 1 \quad ,, \quad ,, \quad ,, \quad ,, \quad ,, \quad ,, \quad ,, \quad \frac{g}{W} \quad ,,$$

$$\therefore F \quad ,, \quad ,, \quad ,, \quad ,, \quad ,, \quad ,, \quad ,, \quad \frac{Fg}{W} \quad ,,$$

If $a =$ acceleration produced by F lb. wt.

$$a = \frac{Fg}{W}.$$

$$\therefore \quad \frac{F}{W} = \frac{a}{g}.$$

The force F need not necessarily be vertical; it may act in any direction.

Example. To find the acceleration produced in a mass of 12 lb. by a force of 3 lb. wt.

Now
$$\frac{F}{W} = \frac{a}{g}$$

and $\quad F = 3$ lb. wt., $\quad W = 12$ lb. wt., $\quad g = 32$ ft./sec.^{2}

$$\therefore \quad \frac{3}{12} = \frac{a}{32}, \qquad \therefore \quad a = 8 \text{ ft./sec.}$$

Third Law of Motion.

If a body A exerts a force on a body B, the body B always exerts an equal and opposite force on A, or *Action and reaction are equal and opposite.*

When you exert a force on a body it always exerts an equal and opposite force on you. Thus you cannot lift yourself and the chair, on which you are sitting, by pulling on its staves, since the chair pulls back with an equal force. You can lift yourself by pulling on a hanging rope as long as you pull downwards on the rope with a force greater than your weight. Then the rope in turn pulls upwards on you with a force greater than your weight. The upper part of the rope is pulling down on the ceiling, and the building transmits the force to the earth; it is therefore the earth which eventually takes the lift.

Suppose there are two boats alongside in the middle of a lake, and a man in one of them pushes the other away. The boat in which the man is sitting does not remain stationary while the other moves away from it. Both boats move apart, because, when the man pushes on the second boat, the second boat pushes back on him with an equal and opposite force.

The bad-mannered method of pushing a chair back from the table is another example of the law. You push forwards on the table but move backwards, since the table pushes back on you with an equal force. The friction between the table and the ground is too large (as a rule) for your push to move the table; hence your push is communicated to the earth as a whole.

When a man throws up a ball he exerts an upward force on the ball and hence the ball exerts a downward force on him. This downward force could be detected and measured by causing the man to stand on a weighing machine and noting the apparent increase in his weight at the moment he throws up the ball.

Now if a horse pulls on the shafts of a cart, and the shafts pull back with an equal force, how can he move? Newton himself answered this problem in some detail.

Consider the horizontal forces acting on the horse. He wants to move forward; the shafts pull back on him with a force P. He thrusts back on the ground with a force F_1 and the ground exerts an equal forward force on him. He will move forward if $F_1 > P$.

It is clear that the possible magnitude of the force F_1 depends on the friction between the horse's hoofs and the ground. On a slippery road F_1 is sometimes less than P and the horse cannot pull the cart until sand has been scattered to increase F_1.

Now consider the horizontal forces acting on the cart. There is a forward pull P on the shafts due to the horse and a backward force equivalent to a pull F_2, due mainly to the friction between the wheels and the axles. The cart will move forward if $P > F_2$.

The life of Newton.

The laws of motion are only part of Newton's contribution to Physical Science. He is universally recognised as one of the greatest scientists of all time, and for sheer intellectual power his work has never been surpassed.

Newton was born in 1642 (the year which saw the beginning of the Civil War between the Royalists and Roundheads and the death of Galileo), at Woolsthorpe in Lincolnshire. As a boy he went to King's School, Grantham, where his name, carved with his own hands upon a window sill, is still proudly shown to-day. At school he was taught Latin and grammar, and showed few signs of his future genius; indeed, he was considered dull until, having been kicked by a bigger boy who was above him in class, he thrashed the bully, and set to work to beat him in his studies. We are told, however, that he was very mechanically minded and fond of making kites, windmills and model machines. This is of special interest in view of his experimental skill in later years.

At the age of nineteen he entered Trinity College, Cambridge, where he began the study of mathematics and science in which his great discoveries were made. In accordance with the tradition which he founded, Cambridge has maintained to the present day its position as the home of British Science.

While still an undergraduate he discovered the Binomial Theorem in Algebra. Just after he had taken his B.A. degree, he did some classic experiments on the breaking up of white light into colours, and invented a new branch of mathematics known as the Calculus.

At the age of twenty-six he became Lucasian professor of Mathematics, a post which he held until he was fifty-four. During this period his greatest discoveries were made. In 1696 he became Master of the Mint, and gave up his work for Science. He was knighted by Queen Anne in 1705. In 1727, at the age of eighty-five, he died and was buried in Westminster Abbey.

It was customary in Newton's time for the great mathematicians of Europe, having solved a particular problem, perhaps after months of work, to offer it as a challenge to all others. Newton invariably solved such problems within twenty-four hours.

He never sought fame, and many of his discoveries had to be coaxed from him years after they had been made. His chief work, the *Principia* (written, as were all the learned works of his period, in Latin), was published by the persuasion and at the cost of his friend Halley.

Many stories are told of his absentmindedness. On one occasion a friend ate his dinner, and Newton remarked: "Dear me, I thought I had not dined, but I see I have".

On another occasion he is said to have left his guests at dinner to fetch more wine, and when after a long interval he did not return, the guests went to seek him. They found him hard at work in his study, completely oblivious of their presence in his house.

One of his most-quoted sayings is his own criticism of his discoveries: "I know not what the world may think of my labours, but to myself it seems that I have been but as a child playing on

Fig. 183. Isaac Newton.

the sea-shore; now finding some prettier pebble or more beautiful shell than my companions, while the unbounded ocean of truth lay undiscovered before me ".

The Law of Gravitation.

Newton's greatest achievement was the explanation of the complicated and mysterious movements of the stars and planets as a deduction from a single simple law—the Law of Gravitation.

If on a clear night you observe at intervals the position of prominent stars and constellations in relation to some fixed object on the earth, such as a tree or chimney, you will find that they are moving across the sky, as the sun does in the day time. In fact, the whole of the heavens appears to be slowly revolving about a point very near the pole star. Certain stars, called planets, do not revolve with the rest but gradually change their position sometimes forwards and sometimes backwards, relative to the rest. The moon moves in a path of her own across the sky.

At first men believed that the whole universe was revolving round the earth. The monk Copernicus (1473–1543) realised that the appearance of motion of the dome of the sky could be preserved if the universe were at rest, and the earth were rotating on its axis, a much neater theory.

In order to account for the motion of the planets and the change in the position of the sun in the sky during the seasons, Copernicus revived the theory that the earth and the planets are revolving round the sun. This theory, at first, met with fierce opposition from almost the whole of educated mankind, and was the cause, as we saw in the preceding chapter, of Galileo's trial by the Inquisition.

Nearly a hundred years later Johann Kepler (1571–1630) discovered three famous laws governing the orbits of the planets. These laws were the result of years of arithmetical trial and error with the figures of Tycho Brahé (1545–1601), the Danish astronomer, who had spent almost every night for twenty years observing the position of the moon and the planets in the sky. Briefly, these laws stated that the planets, including the earth, revolved round the sun in flattened circles, called ellipses, and that there was a simple relation between the rate at which they revolved and their distance from the sun.

This was the state of affairs when Newton applied his mind to the problem. He asked himself the question: "Why do the planets revolve in this way?" When a body moves on the earth we always look for some cause, but no one, before Newton's time, had been able to give a satisfactory explanation, other than supernatural, of the movement of the heavenly bodies.

We are told that one day Newton was sitting in his orchard at Woolsthorpe, when he observed an apple fall to the ground. He reasoned that the apple fell because the earth was attracting it. If you take the apple to the summit of a high mountain, the earth still attracts it. How far away would the apple need to be taken before the earth ceased to attract it? Was the moon, wondered Newton, too far away for the earth to attract it? It occurred to him that the moon might be revolving round the earth because the earth was attracting it, just as a stone can be made to revolve round the head of a boy by his constant pulling on a string.

Before he could test the hypothesis he had to make a guess at the law which governed the decrease in the attraction of the earth with increasing distance from it. He considered an inverse square law most probable $\left(\text{Force} \propto \dfrac{1}{(\text{Distance})^2}\right)$, and on this assumption he calculated what the speed of the moon would be. The result differed from the true value, 1 revolution in $27\frac{1}{3}$ days, by 18 per cent. Newton considered the discrepancy too great and put his calculations in a drawer, where they remained for sixteen years. This is a wonderful instance of intellectual honesty.

In 1682, a more accurate measurement of the size of the earth (an important quantity in Newton's calculations) had been made, and with the new figures he worked out his calculation again. When he saw that the result was likely to come out correctly it is said that he was seized with such nervous excitement that he had to ask a friend to finish the calculation.

Thus Newton proved his theory of universal gravitation, the theory that all the bodies in the universe are attracting each other with a force which diminishes as the inverse square of the distance. With superb mathematical skill he applied it to the motion of the planets, the perturbation of the moon's orbit owing to the attraction of the sun, the tides and many other problems.

Gravity—the attraction of the earth for bodies—is a particular example of gravitation. We do not notice the attraction of terrestrial objects for each other because it is so small in comparison with the pull of that enormous body, the earth, on each of them. In other words, their mutual attractions are negligible compared with their weights. However, it is found that a plumb-line may be pulled slightly out of the vertical by the sideways attraction on its bob of a large body, such as a mountain. The experiment was performed on the slopes of the mountain, Schiehallion, in Scotland.

Whenever a simple fundamental law like the law of gravitation is discovered it opens up a new chapter of science. Mathematicians apply it to solve more complicated problems; they explore its consequences and sometimes make the most unexpected deductions.

An application of Newton's theory which caused a great sensation was the discovery, in 1846, many years after his death, of the planet Neptune. Some force had been pulling Uranus out of its path. Two mathematicians, Adams and Leverrier, independently assumed the cause to be an unknown planet, and calculated its requisite position. Astronomers turned their telescopes to the predicted spot in the sky, and found the planet.

Although it was not such a piece of mathematical skill, an even greater sensation was caused by the successful fulfilment of Halley's prophecy that the comet which bears his name would return (long after his death) in 1759.

The Nautical Almanac, published four years in advance, giving the dates of the phases of the moon, the times of the tides, predicting the exact times and positions of eclipses, the positions of the planets and their satellites, is another impressive application of the law of gravitation.

Newton's demonstration that even the motions of the heavenly bodies are subject to simple laws has been a great inspiration to the men of science who have followed after him. It is related that Napoleon, to whom the famous French mathematician, Laplace, had been explaining his theory of celestial mechanics, remarked, "But you have not mentioned God in your theory". "Sire", replied Laplace, "I have no need of that hypothesis". Science endeavours to give a natural explanation of all material phenomena. Anything which is incapable of such an explanation lies outside its scope.

Experiments with a pendulum.

A simple pendulum consists of a small "weight", called a bob, suspended by a fine thread. It has a remarkable and useful property which was discovered by Galileo. He is said to have timed by means of his pulse the swing of a chandelier hanging in Pisa cathedral. He found that each complete swing occupied a constant interval of time. He therefore concluded that the time of swing of a pendulum is independent of the extent of the swing. This has been proved to be correct so long as the swing is a small one, the angle not exceeding about 20° on either side of the vertical. A pendulum is therefore said to be isochronous (from the Greek, meaning "equal time").

Experiment 1. To show that the time of swing of a pendulum is independent of the weight of the bob. Take two pendulums, of exactly the same length from the point of suspension to the centre of gravity of the bob, the one having a bob of wood, and the other a bob of lead of approximately the same size. Time their rate of swing.

(If a whole form is doing the experiment it is convenient for some boys to use wooden and others leaden bobs, and then their results may be compared.)

Attach the pendulums to a rigid support by clamping the thread between two pieces of wood. If the thread is looped round something, such as a pencil, its length will alter slightly as it swings. A length of 100 cm. is suggested.

For accurate timing a stop-watch is not essential; an ordinary watch with a seconds finger may be used. Take the time for 50 complete swings to and fro, and repeat the experiment twice so that three sets of readings are obtained. Record all these readings since their consistency gives a good indication of the accuracy of the experiment. It is convenient for one boy to read the watch and another boy to count. The boy counting should give the signal for times to be taken as the pendulum is passing through its vertical position (a vertical line may be drawn behind the pendulum to assist him), since a much sharper end point may be obtained in this position than at the end of a swing.

Take care in starting the pendulum that it swings in a vertical plane, not in an ellipse which it will readily do, and see that the swings are small.

A deduction from the experiment. A pendulum swings because the weight of the bob moves its mass; its time of swing depends on both the mass and the weight of the bob. Since the time of swing is the same with different bobs, we may deduce that the weights of the bobs are proportional to their masses.

If a pendulum is taken to the top of a high mountain, the weight of its bob diminishes, since it is farther from the centre of the earth, but the mass of the bob is unchanged. The pendulum, therefore, swings more slowly.

The fact that the weights of all bodies are proportional to their masses at a particular place may be deduced from Galileo's experimental demonstration that all bodies fall with the same acceleration, but the pendulum method is a much more delicate proof.

Experiment 2. To find how the time of swing of a pendulum depends on its length. Repeat Experiment 1 for several lengths of the pendulum, shortening it from about 100 cm. to 30 cm., by about 10 cm. at a time. Measure the length from the point of support to the centre of gravity of the bob.

Obtain the average time for one swing, T, for each length, l, of the pendulum. Plot T against l, and T^2 against l.

What do you deduce from your graphs?

The use of a pendulum in a clock.

The isochronous property of a pendulum makes it admirably suited for regulating a clock. But how can a pendulum be driven by and yet control a clock? The answer is that the "free" time of swing of the pendulum is always slightly affected by the driving mechanism, but in the best modern clocks the interference is very slight.

The pendulum controls the clock and is kept swinging by what is known as an escapement mechanism. An anchor-shaped pallet BAC (see Figs. 184 (*a*) and (*b*)) is mounted so as to vibrate with the pendulum. The pallet engages with the teeth of the escape wheel D which is driven by the clock spring, and tries to turn in the direction of the arrow. When the pendulum is at the extreme left of its swing, it knocks B out and C in, and D goes round through one tooth. Thus at each half swing (either to or fro) of the pendulum, D goes round through one tooth. The seconds finger of the clock is geared to D and can be seen to move in a series of jerks which occur every time the latter moves round through one tooth.

Thus D is regulated by the pendulum. The swings of the pendulum are prevented from dying down by shaping the teeth of D and the edges of the pallet at B and C so that an impulse is

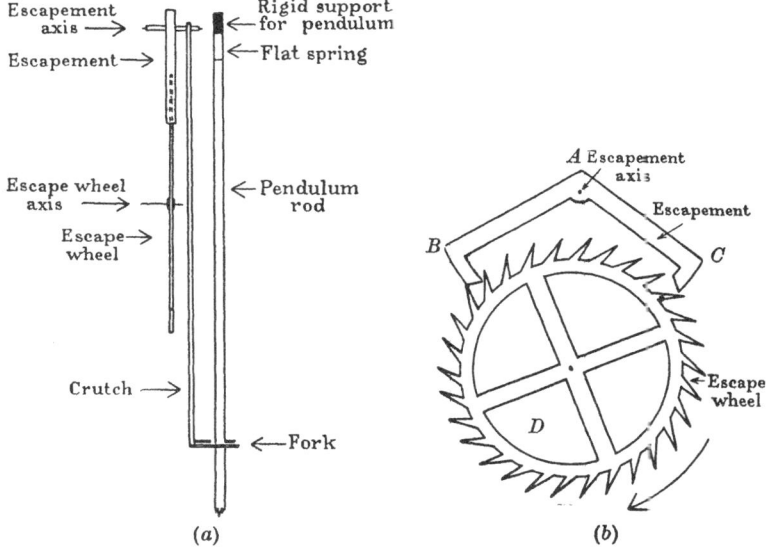

Fig. 184. Escapement mechanism of a clock.

given to the pendulum by the fork each time a tooth is sliding on the pallet edge while "escaping".

A watch is controlled by a balance wheel instead of a pendulum, but its escapement mechanism is similar to that of a clock. You should open your watch and look at the escapement mechanism working.

SUMMARY

Newton's Laws of Motion.

1. A body continues in a state of rest or to move with a steady velocity in a straight line if it is not acted upon by forces.

2. When a force acts on a body it produces an acceleration which is proportional to the magnitude of the force:

$$\frac{F}{W} = \frac{a}{g}.$$

(This law is not complete: see p. 255.)

3. If a body A exerts a force on a body B, the body B always exerts an equal and opposite force on A, *or* Action and reaction are equal and opposite.

Newton's Law of Gravitation.

Every body in the universe is attracting every other body with a force F which depends on their distance apart, d:

$$F \propto \frac{1}{d^2}.$$

Weight is a particular example of the force of gravitation.

The simple pendulum.

The time of swing of a pendulum is independent of the extent of the swing (as long as the latter is small), and is proportional to the square root of the length of the pendulum.

QUESTIONS

1. Explain the following carefully:

(a) When a corridor train starts or stops suddenly, a sliding door may open or shut.

(b) A cyclist tends to fall outwards and not inwards when rounding a corner too sharply.

(c) A circus rider jumping through a hoop jumps vertically. What would happen if he took a vigorous leap forward?

(d) When you step ashore from a rowing boat it tends to leave the shore.

(e) A penny dropped in a train travelling at 60 m.p.h. is not dashed against the side of the carriage.

2. State Newton's First Law of Motion.

Account for the fact that the engine of a car, travelling on the level, may be pulling, and yet the speedometer registers a steady velocity.

3. State Newton's Third Law of Motion.

In a tug of war, for one side to exert a pull on the rope, the other side must pull back with an equal and opposite force. How then can one side pull the other over? Explain fully.

4. In Question 3, if the teams are pulling on uniform concrete with similar shoes, is it muscular strength or weight which tells most in deciding which side will win?

5. What force is necessary to give a body of weight 10 lb. a horizontal acceleration of 6 ft. per sec. per sec., ignoring friction?

6. A boy of weight 10 stone is skating and the friction between him and the ice is 10 lb. What is his retardation (i.e. negative acceleration) when, having got up speed, he allows himself to glide to rest?

7. What force must an engine exert on a train of weight 500 tons in order to give it an acceleration of $\frac{1}{10}$ ft. per sec. per sec., taking the friction as 14 lb. wt. per ton? (Find the force necessary to accelerate the train, ignoring friction, and then add to this the force necessary to overcome friction.)

8. A simple pendulum hangs in a train. What happens to it when the train is (a) starting, (b) travelling at a high uniform speed, (c) slowing down?

Explain fully why it behaves in the way you suggest. [This device is used to indicate and measure acceleration.]

9. A man stands on a weighing machine in a lift. Describe and explain how the reading of the machine changes as the lift ascends from start to stop.

10. A body of weight 2 lb. is suspended from a spring balance in a lift. Find the reading of the spring balance when the lift is (a) ascending, (b) descending, with an acceleration of 4 ft./sec.^{2} (Take $g = 32$ ft./sec.^{2})

School Certificate Questions

11. A stone thrown with a velocity of 10 ft. per sec. along a horizontal sheet of ice comes to rest in 100 ft. What is the ratio between the frictional force and the weight of the body?

12. State Newton's laws of motion and explain the meaning of the terms you use. Assuming the acceleration due to gravity to be six times as great at the surface of the earth as at the surface of the moon, compare the forces required to impart an acceleration g to a bullet at the two places (a) in a horizontal direction, (b) in a direction vertically upwards.

13. Explain carefully the distinction between weight and mass. How would you show experimentally that the two quantities are not the same? If you were a grocer on the moon, would you find it more profitable to use a spring balance manufactured in Birmingham or a pair of scales and set of weights of similar origin?

Chapter XII

FORCE AND MOTION

Newton's Second Law of Motion.

We have already explained (see p. 241) that when a force acts on a body it produces not a velocity, but a change of velocity, i.e. it produces an acceleration. According to Newton's second law of motion, this acceleration is proportional to the force. But when a force acts on a body it is clear that the acceleration produced depends also on the body's mass—the greater the mass the smaller the acceleration. It is found, in fact, that if the mass is doubled, the acceleration is halved, and speaking generally, the acceleration produced by a given force is inversely proportional to the mass.

We may express these facts mathematically as follows:

1. $a \propto F$,

2. $a \propto \dfrac{1}{m}$,

where a = acceleration, F = force, m = mass.

$$\therefore a \propto \frac{F}{m},$$

$$F \propto ma,$$

$$F = kma,$$

where k is a constant.

By choosing a new unit of force, we can make $k = 1$. The equation then becomes

$$\mathbf{F = ma.}$$

On the British system this unit of force is the poundal.

1 poundal is the force required to give a mass of 1 lb. an acceleration of 1 ft. per sec. per sec. (When $m = 1$, $a = 1$, $F = 1$, and therefore $k = 1$.) Thus in the equation $F = ma$, if m is in lb., a in ft. per sec. per sec., then F will be in poundals.

Since the weight of 1 lb. produces in a mass of 1 lb. an acceleration of 32 ft. per sec. per sec.,

$$1 \text{ lb. wt.} = 32 \text{ poundals.}$$

Speaking generally, the weight of a body of mass m lb. $= mg$ poundals.

On the Metric system, the unit of force, the dyne, is based on the cm., gm., and sec., and this system is, in consequence, known as the C.G.S. system of units.

1 dyne is the force required to give a mass of 1 gm. an acceleration of 1 cm. per sec. per sec.

$$1 \text{ gm. wt.} = 981 \text{ dynes.}$$

The poundal and the dyne are known as absolute units. Unlike the lb. wt. and the gm. wt. (called gravitational units) they are invariable and entirely independent of geographical position.

We can now make a full statement of **Newton's Second Law of Motion**: *When a force acts upon a body, it produces an acceleration which is proportional to the force and inversely proportional to the mass of the body*, i.e. $F = ma$.

This is the most fundamental fact of Dynamics, which is mainly the study of the effect of forces in producing motion, i.e. the relation between force, mass and motion. While it is capable of experimental proof with suitably designed apparatus, our ultimate belief in its truth lies in the fact that it always " works " when applied to problems—whether it be the motion of a bullet or a comet that it is used to calculate.

Indeed, this is a common procedure in science. A fundamental law of widespread application is discovered as an inspired guess. It is then applied to solve problems, the solutions to which are capable of experimental verification. In each case the theoretical result is compared with that obtained from experiment, to see whether they are in accord. Thus a mass of indirect evidence is collected, until, as in the case of Newton's second law of motion, its testimony is overwhelmingly in favour of the law.

Fletcher's trolley.

An apparatus known as Fletcher's trolley (see Fig. 185) has been specially designed to demonstrate experimentally the truth of Newton's second law of motion. It may be used to prove that

1. *Using a body of fixed mass, the acceleration is directly proportional to the force applied* ($a \propto F$).

2. *Using a given force, the acceleration of a body is inversely proportional to its mass* $\left(a \propto \dfrac{1}{m}\right)$.

The "body" used is a trolley whose wheels run in grooves in a steel plane, and whose mass may be increased by inserting steel cylinders.

The force is applied by weights hung on a string attached to the pulley and passing over a pulley at the end of the plane.

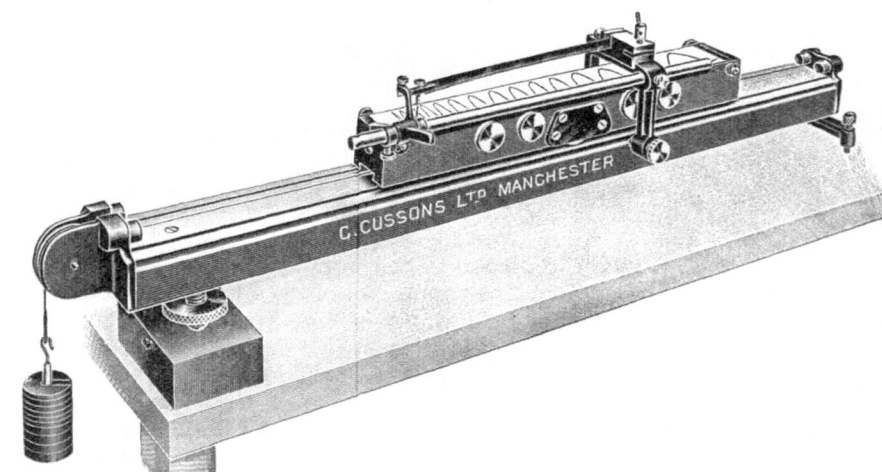

Fig. 185. Fletcher's Trolley.

The acceleration is measured by means of a steel strip, of known time of vibration, and carrying an inked brush, which makes a wavy trace on a strip of paper pinned to the top of the trolley.

Friction is first counteracted by tilting the plane slightly by means of the levelling screws so that the trolley runs down the plane with uniform velocity when given a start. The lengths of the waves made by the brush will then all be equal.

Different suitable weights are hung on the end of the string and a wavy trace obtained in each case (keeping the mass of the trolley constant).

To find the acceleration, a central line is drawn on the wavy trace and the length of each complete wave measured. These lengths represent the distances traversed by the trolley in equal time intervals.

$$\therefore \quad \text{Average velocity during each wave} = \frac{\text{Length of wave}}{\text{Time interval}}.$$

Subtract the average velocity in each wave from that in its successor. This gives the gain of velocity per unit time interval.

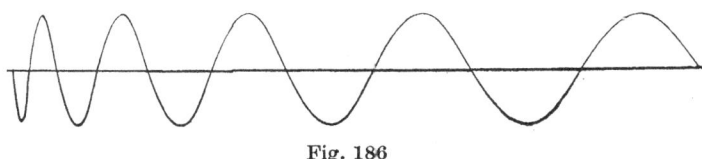

Fig. 186

Hence the gain of velocity per unit time, i.e. the acceleration, may be found.

Two sets of values are given in the table below.

Unit of time taken = time for 1 complete oscillation of steel strip (instead of 1 sec.).

Force pulling trolley (lb. wt.)	Successive wave-lengths (in.) = Average Velocity in interval (in. per time unit)	Increase in Velocity in each time interval = Acceleration	Average Acceleration	Acceleration / Force
$\frac{4}{10}$	2·53 3·38 4·24 5·08	0·85 0·86 0·84	0·85	2·1
$\frac{5}{10}$	2·62 3·71 4·75 5·81	1·09 1·04 1·06	1·06	2·1

It will be seen from the table that the acceleration in each case is fairly constant and that the ratio $\dfrac{\text{acceleration}}{\text{force}}$ is constant, proving that the acceleration is proportional to the force.

The second part of the experiment is performed by repeating the procedure, keeping the force constant and varying the mass of the trolley.

It should be noted that the tension in the string is not exactly equal to the hanging weight when the latter is moving. For the "weight" at the end of the string would not move unless its weight were greater than the tension of the string. However, the difference is not large if the trolley is considerably heavier than the hanging weight.

Absolute C.G.S. units of work and power.

Units of work

1 erg = work done when 1 dyne acts through 1 cm.

1 joule = 10^7 ergs.

Unit of power (or rate of working)

1 watt = 1 joule per sec.

The c.g.s. system of units pervades the whole of physics, giving a most satisfying coherence. We shall see, for instance, that the units used in magnetism and electricity are also, fundamentally, c.g.s. units.

Expression for kinetic energy, $\frac{1}{2}mv^2$.

Often in solving a problem dealing with the motion of a body it is convenient to consider the changes in its kinetic energy (see p. 185). We shall therefore investigate how the kinetic energy of a moving body depends on its mass and its velocity.

Consider a body of mass m lb. moving with a velocity v ft. per sec. Suppose a constant force of F poundals will bring it to rest in a distance of s ft. By the principle of the conservation of energy (see p. 185) the work done against the force is equal to the kinetic energy of the body.

Let a ft./sec.^{2} = Retardation of the body.

Now, since the final velocity is 0,

$$v^2 = 2as \quad \text{(see p. 233)}.$$

$$\therefore \quad a = \frac{v^2}{2s}.$$

But $\qquad\qquad F = ma$ (Newton's second law).

$$\therefore\ Fs = mas$$

$$= m \cdot \frac{v^2}{2s} \cdot s$$

$$= \tfrac{1}{2}mv^2 \text{ ft. poundals.}$$

But $\qquad\qquad Fs = $ Work done against the force.

$\therefore$ **Kinetic energy of the body $= \frac{1}{2}$mv² foot poundals.**

If m is in gm. and v in cm. per sec., then the kinetic energy $\frac{1}{2}mv^2$ is in ergs.

The Principle of the Conservation of Energy.

We have made use of the Principle of the Conservation of Energy in the preceding paragraph. We will now express it in the form of a general mathematical equation. Suppose a moving body is acted upon by a force causing it to change its velocity.

Let $\quad F = $ Force acting on the body (poundals),

$\qquad s = $ Distance gone (ft.),

$\qquad u = $ Original velocity of the body (ft. per sec.),

$\qquad v = $ Final velocity of the body (ft. per sec.),

$\qquad m = $ Mass of the body (lb.).

$\qquad$ **Fs $\qquad\qquad = \frac{1}{2}$mv² − $\frac{1}{2}$mu²,**

$\qquad$ **Work done = Change of kinetic energy.**

Example. A car of mass 1600 lb. is braked and slows down from 30 m.p.h. to 15 m.p.h. in 10 yd. What retarding force do the brakes exert on the car?

$$30 \text{ m.p.h.} \quad = \frac{30 \times 1760 \times 3}{60 \times 60} = 44 \text{ ft. per sec.}$$

$$15 \text{ m.p.h.} \quad = 22 \text{ ft. per sec.}$$

Let $\quad F$ poundals $=$ Retarding force of brakes.

Now

Work done *by* car $=$ Loss of kinetic energy.

$$\therefore \quad F \times 10 \times 3 = (\tfrac{1}{2} \times 1600 \times 44^2) - (\tfrac{1}{2} \times 1600 \times 22^2),$$

$$F = \frac{\tfrac{1}{2} \times 1600}{10 \times 3} \cdot (44^2 - 22^2)$$

$$= 38{,}920 \text{ poundals}$$

$$= \frac{38{,}920}{32} \text{ lb. wt.}$$

$$= \frac{38{,}920}{32 \times 2240} \text{ ton wt.}$$

$$= 0 \cdot 540 \text{ ton wt.}$$

Momentum.

In the seventeenth century, a long controversy developed between two mathematicians, Leibnitz and Descartes, concerning the "efficacy" of moving bodies. Now "efficacy" is a vague term, and the dispute arose because each attached to it a different meaning. To Leibnitz the efficacy of a moving body was its kinetic energy, or its capacity to do work. He maintained that it was proportional to the square of the velocity of the body. We know now that he was right, because the kinetic energy of a moving body with mass m and velocity v is $\tfrac{1}{2}mv^2$.

Descartes, on the other hand, said that the efficacy of a moving body was proportional to its velocity. He was equally right, for he was thinking of what is now known as the **Momentum** of the body, mv.

What precisely does the momentum, i.e. the product of the mass and the velocity of a body, represent? It represents the quantity of motion that a force can impart to a body in a given time. This again is a vague statement, but we will express it in a mathematical equation.

Suppose a body of mass m is travelling at a velocity u, when a force F acts on it for a time t and changes its velocity to v. Let a be the acceleration of the body while the force is acting.

Now $\qquad\qquad\qquad F = ma.$

But $\qquad\qquad\qquad v = u + at$ (see p. 232).

$$\therefore \quad a = \frac{v - u}{t}.$$

Substituting for a,

$$F = m \frac{v-u}{t},$$

i.e. $Ft = mv - mu,$

Impulse = Change of momentum.

Thus the product of force and time, which is called the *impulse*, is equal to the change of momentum. If m is in lb. and v in ft. per sec., the units of both momentum and impulse are poundal-sec. On the c.g.s. system, the corresponding units are dyne-sec.

The momentum equation is particularly useful in solving problems dealing with collision, or forces acting for a very short time, such as the blow of a bat on a ball.

Example. A cricket ball weighing $5\frac{1}{2}$ oz. and moving at 60 ft. per sec. is brought to rest in $\frac{1}{4}$ sec. by the hands of a fieldsman. What is the average force it exerts on his hands?

Let F poundals = Average force ball exerts.
Now Ft = Change of momentum.
∴ $F \times \frac{1}{4} = \frac{5\frac{1}{2} \times 60}{16},$
            $F = 82\frac{1}{2}$ poundals
            $= \frac{82\frac{1}{2}}{32} = 2 \cdot 58$ lb. wt.

Principle of the Conservation of Momentum.

When two bodies such as two billiard balls collide, there is no loss or change of momentum.

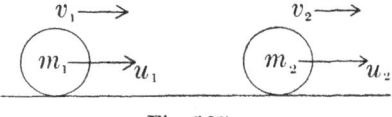

Fig. 187

Momentum before impact = Momentum after impact,
$$m_1 u_1 + m_2 u_2 = m_1 v_1 + m_2 v_2,$$
where m_1, m_2 represent the masses of the balls, u_1, u_2 their original velocities, and v_1, v_2 their velocities after impact. All the velocities must be taken in the same direction. If they are in the

opposite direction they must be regarded as negative. For momentum is a vector quantity, and we must always take account of its direction.

The principle of the conservation of momentum follows direct from Newton's third law of motion, "Action and reaction are equal and opposite". Consider the two billiard balls. They must exert equal and opposite forces on each other, and since the time of contact is the same for both, the impulses must be equal and opposite. Consequently their changes of momentum must be equal and opposite, thus cancelling out. Hence there is no net change of momentum.

The principle of the conservation of momentum is capable of a wider application than simple collision. For instance, when a gun recoils after firing a shot, the backward momentum of the gun is equal to the forward momentum of the shot. This is due to the fact that the exploding charge exerts the same force on both shot and gun for the same time. The same applies to the recoil of a hose-pipe. Again, the rocket car moves forward by ejecting gases from pipes in its rear just as a gun recoils, and the forward momentum of the car is equal to the backward momentum of the gases.

We may state the principle in general terms as follows: **When two or more bodies act on each other (provided no external force acts on the whole system), there is no total change of momentum.**

Thus the centre of gravity of a shell which has burst in midair continues to move in its normal path, since no external force has acted upon the system. The bits fly in all directions, but their centre of gravity moves on as before.

Loss of kinetic energy on collision.

Although there is no loss of momentum when two bodies collide, there is always a loss of kinetic energy (usually converted into heat). This seeming paradox is due to the fact that momentum is a vector quantity, whereas kinetic energy is not. Thus if two express trains of equal weight are approaching each other from opposite directions at 60 m.p.h. their momenta are equal and opposite. Thus their total momentum both before and after collision is zero. Their total kinetic energy before impact, however, is considerable, and all this is lost or wasted during the collision, in smashing up the locomotives and coaches and up-rooting the track.

Example 1. *A locomotive of mass* 20 *tons, travelling at* 10 *ft. per sec., collides with a loaded truck of mass* 5 *tons and the two are automatically coupled, and move on together. Find their common velocity, and the loss of kinetic energy at collision.*

Let v ft. per sec. = Common velocity after collision.

Momentum after impact = Momentum before impact.

$$\therefore \quad (20+5)\, v = 20 \times 10.$$

$$\therefore \qquad v = \tfrac{200}{25} = 8 \text{ ft. per sec.}$$

Kinetic energy before impact $= \tfrac{1}{2} m v^2$

$$= \tfrac{1}{2}.20.10^2 \text{ ft. tondals}$$

$$= \tfrac{1000}{32} = 31\tfrac{1}{4} \text{ ft. tons.}$$

Kinetic energy after impact $= \tfrac{1}{2}.25.8^2 = 800$ ft. tondals

$$= \tfrac{800}{32} = 25 \text{ ft. tons.}$$

$\therefore$ Loss of kinetic energy at collision

$$= 31\tfrac{1}{4} - 25$$

$$= 6\tfrac{1}{4} \text{ ft. tons.}$$

Example 2. *A* 13·5 *in. gun has a mass of* 67 *tons, and gives a shell of mass* 1250 *lb. a velocity of* 2000 *ft. per sec. Find the velocity of recoil of the gun.*

Let v ft. per sec. = Velocity of recoil of gun.

Backward momentum of gun = Forward momentum of shell.

$$\therefore \quad 67 \times 2240 v = 1250 \times 2000.$$

$$\therefore \qquad v = \frac{1250 \times 2000}{67 \times 2240}$$

$$= 16 \cdot 65 \text{ ft. per sec.}$$

The gun is brought to rest by means of hydraulic buffers.

SUMMARY

Newton's three laws of motion are adequate for dealing with any problem in dynamics, i.e. a problem on the motion of bodies and the forces involved. However, it is convenient for solving special types of problems, to use the energy and momentum equations, which can be deduced from Newton's second law.

The following three equations summarise almost the whole of dynamics.

1. Newton's Second Law of Motion:

$$F = ma.$$

2. Principle of the Conservation of Energy:

$$Fs = \tfrac{1}{2}mv^2 - \tfrac{1}{2}mu^2,$$

Work done = Change of kinetic energy.

3. Momentum equation:

$$Ft = mv - mu,$$

Impulse = Change of momentum.

From equation (3) the Principle of the Conservation of Momentum follows direct: "When two or more bodies act on each other (provided no external force acts on the whole system), there is no total change of momentum".

Units.

In the above equations, using lb. ft. sec. or cm. gm. sec. units,

F is in poundals or dynes,

$\tfrac{1}{2}mv^2$ is in ft. poundals or ergs,

mv is in poundal secs. or dyne secs.

QUESTIONS

(mainly from School Certificate Papers)

1. Consider a bullet of mass 1 oz. and velocity 2000 ft. per sec., and also a steam roller of mass 5 tons and velocity 3 ft. per sec.

(i) Find (a) the momentum, (b) the kinetic energy, of each.

(ii) Calculate (a) the time required, (b) the distance required, for a force of 1000 poundals to stop each of them.

Discuss the significance of your answers.

2. A railway truck of mass 10 tons moving at 6 m.p.h. hits another truck of mass 8 tons and the two move on together. What is their common speed? How much kinetic energy was lost during the collision?

3. (a) A gun has a mass of 29 tons and gives to a projectile, mass 500 lb., a velocity of 2040 ft. per sec. Find the velocity of recoil of the gun.

FORCE AND MOTION 265

(b) Find the kinetic energy of the shell and the gun when the former has just left the barrel?

Since an equal force has acted on both, how do you account for the difference?

4. A machine gun fires 300 1 oz. bullets per minute, giving each a velocity of 1600 ft. per sec. What force is necessary to hold it still?

5. Two equal masses of 200 gm. are hung from the ends of a string passing over a pulley. A mass of 10 gm. is placed on one of them, and it descends a distance of 103·5 cm. in exactly 3 sec. Find its acceleration.

Taking 410 gm. as the total mass moved and 10g dynes as the motive force, obtain an equation and find g. Can you suggest why the value of g is too small? (This is known as Atwood's experiment.)

6. State Newton's Second Law of Motion and describe in detail an experiment by which it may be verified in the laboratory.

7. What is the relation between force, mass and acceleration?

The cage of a lift has a mass of 500 kilograms. What is the pull in the rope to which the cage is attached (a) when the lift is descending with uniform velocity, (b) when it is descending with an acceleration of 40 cm. per sec. per sec.?

8. An engine of mass 20 tons is travelling at a constant velocity of 30 m.p.h. along a smooth level track. What steady force can bring it to rest (a) in ½ min. (b) in ½ mile?

9. When a man springs ashore from a boat, the boat is simultaneously driven away from the bank. Why is this? If the boat and man were originally at rest, the man weighing 140 lb. and the boat 200 lb., and the man jumped ashore with a velocity of 10 ft. per sec., what would be the velocity of recoil of the boat if the resistance of the water be negligible?

10. The shell fired from a gun of 12 in. bore and 40 ft. in length weighs 880 lb., and attains a muzzle velocity of 2400 ft. per sec. What uniform pressure in the barrel will cause this velocity? ($\pi = \frac{2\,2}{7}$.)

11. A tramcar of mass 5 tons starts from rest with uniform acceleration and in 1½ min. attains a speed of 20 m.p.h. Find (a) the acceleration, (b) the momentum 1 min. after starting, (c) the kinetic energy 1½ min. after starting.

12. A train whose total weight is 300 tons is running with the uniform velocity of 40 m.p.h. when the brakes are applied, reducing the velocity to 20 m.p.h. in half a mile. What energy has the train lost and what has been the average retarding force employed?

13. A motor car weighing $1\frac{1}{2}$ tons and travelling at 15 m.p.h. shuts off its engine just at the beginning of an incline which rises 1 ft. vertically for every 24 ft. length of road.

(a) Find the distance the car will proceed along the slope before coming to a standstill.

(b) Find the loss in kinetic energy when it has travelled 120 ft. along the incline.

14. State the Principle of the Conservation of Energy, and show that for a body falling freely under gravity the sum of its potential and kinetic energies is constant.

A body (of mass m) slides from rest down a plane inclined at an angle of 30° to the horizontal. After sliding a distance of 20 ft. down the plane, it is found to be moving with a velocity of 16 ft. per sec. How much energy has been lost in overcoming friction?

15. Explain the meaning of work, energy, power.

A fire engine pumps water from a well 25 ft. deep, and projects it with a velocity of 72 ft. per sec. Find the horse-power developed by the engine if the water is delivered at the rate of 600 gallons per minute. (1 gallon of water weighs 10 lb. 1 H.P. = 550 ft. lb. per sec.)

16. Water is projected perpendicularly with a velocity of 50 ft. per sec. against a wall 3 ft. square. Assuming that 150 gallons of water are projected per min. and that there is no splashing back, calculate the mean pressure on the wall. (A gallon of water weighs 10 lb.)

17. A bullet weighing $\frac{1}{2}$ oz. has its speed reduced from 1000 ft. per sec. to 200 ft. per sec. in passing through a wooden board 3 in. thick. What is the average resistance offered to the bullet's motion and what is its loss of kinetic energy in passing through?

18. Show that the kinetic energy of a body of mass m moving with a velocity v is measured by $\frac{1}{2}mv^2$.

A bullet weighing 1 oz. is fired with a velocity of 1200 ft. per sec. horizontally into a block of wood weighing 1 lb. suspended by vertical strings. The bullet remains embedded in the wood. How much kinetic energy is lost in the impact? What becomes of the energy so lost?

19. A box of sand weighing 10 kilograms hangs by a long light rope. When a bullet weighing 45 gm. moving horizontally strikes the box and remains buried in it, the box swings back until its centre of gravity has risen 15 cm. above its initial level. At what speed is the bullet travelling when it strikes the box?

20. The diagram shows a switchback railway. A car weighing 200 lb., starting from rest at *A*, is found to have a velocity 40 ft. per sec. at *B* and comes to rest after rising to *C*. The length of rail between *A*

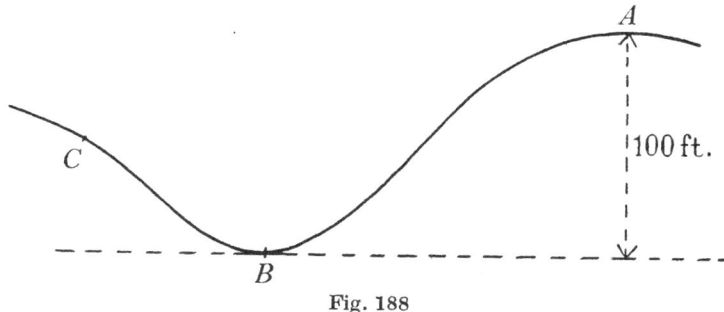

Fig. 188

and *B* is 1000 ft., and between *B* and *C* is 200 ft. If *A* is 100 ft. above *B*, calculate (*a*) the kinetic energy of the car at *B*, (*b*) the work done against the friction of the rails between *A* and *B*, (*c*) how high *C* is above *B*.

Answers to Questions

CHAPTER II (page 23)

11. (a) 1693 gm.; (b) 4704 gm.   12. 208 gm.
16. 1·08 gm. per c.c.   17. 20 c.c.   18. 50 c.c.
19. 70 gm.   20. 2·5 gm. per c.c.
21. (a) 10 c.c.; (b) 105 gm.

CHAPTER III (page 61)

3. (a) 2040 gm. wt.; (b) 68 gm. wt. per sq. cm.   4. 172·8 ft.
5. 16·8 sq. in.   7. 2·24 ft.   8. (a) 6697 tons wt.
13. 20,040,000 dynes per sq. cm.   15. 1 kgm.
16. 3070 gm. wt. per sq. cm.

CHAPTER IV (page 92)

6. 2510 lb. wt.   7. 11·5 metres.   12. 2560 cu. in.
13. 75·6 cm. of mercury.   14. 26·6 lb. per sq. in.   15. 422 metres.
19. (a) $\frac{8}{9}$ atmos.; (b) $\frac{9}{720}$ atmos.   20. 22·1 cm.
21. 93 metres.   22. 20·25 metres.   23. 75·9 cm.

CHAPTER V (page 113)

4. 1·65 cu. ft.   5. 7·56 gm. per c.c., 29·12 c.c.
6. (a) 2·7 gm. per c.c.; (b) 0·8 gm. per c.c.   7. 109·7 gm.
8. 10·06 gm.   9. $10\frac{5}{7}$ tons.   10. $\frac{6}{10}$.
11. 2,100,000 cu. ft.   12. 0·25 gm. per c.c.   13. 72 lb. wt.
14. 1·35 c.c., 3·38 gm.   15. $\frac{3}{4}$.   16. 12·4 tons wt.
17. 1770, 1210, 560 cu. ft.   19. Increase of 10,500 lb. wt.
20. 2·5.   21. 1·45 gm. per c.c.   22. 21·3 c.c., 2·37 cm.
23. 24 gm.   24. 220 gm.   25. 1·05 gm. per c.c.
26. 10·9 cm.   27. 9.

CHAPTER VI (page 144)

5. 1 ft.   **6.** 12 lb. wt., 5.   **7.** $53\frac{1}{3}$ lb. wt., $\frac{1}{3}$.
8. $3\frac{1}{3}$ lb. wt.   **9.** 5·3 ft. from pivot.   **10.** 4, 12, 16 lb. wt.
11. 40·252, 39·750 gm.   **13.** 325, 0, 162·5 lb. in.
14. $1\frac{1}{2}$ tons.   **16.** 120 gm.   **17.** $26\frac{2}{3}$, $33\frac{1}{3}$ tons wt.
19. 315 and 1050 lb. wt. per ft. length of wall. It will topple.
20. 130 lb. wt.   **22.** $4\frac{2}{3}$ ft. from man's end.
23. $126\frac{2}{3}$, $253\frac{1}{3}$ lb. wt.   **24.** (a) 625 lb. wt.; (b) $104\frac{1}{6}$ lb. wt.
26. 150 lb., 3·73 ft. from A.   **28.** 26° 34'.
30. 0·465 in. from centre of rectangle.   **31.** $\frac{8}{9}$ in. from centre.
32. 0·014 of a side.   **33.** 28 in. above level of board, 20° 33'.
34. $16\frac{2}{3}$ gm.

CHAPTER VII (page 176)

5. $\frac{1}{4}$, 80 %.   **6.** 26, 23·9 %.   **7.** 8 ft.
8. 113, 66·1 lb. wt.   **10.** 4, $31\frac{1}{4}$ lb. wt.   **11.** 30·72 lb. wt., 576 ft. lb.
12. 12 tons wt.   **13.** 8, 2 lb. 3 oz., 6·4, 80 %.   **14.** 90 %.
15. 50, 157, 31·8 %.   **16.** 39 lb. wt.

CHAPTER VIII (page 189)

1. 611 H.P.   **2.** 1075 H.P.   **3.** (a) 4400 ft. lb.; (b) 1100 ft. lb.
6. 14·2 H.P.   **7.** 1562·5 lb. wt.   **8.** 4280 ft. lb., 61·1 %.
9. 1·04 tons.   **11.** 240,000 ft. lb., 8·0 H.P.   **14.** 320 H.P.
15. 11·2 H.P.   **16.** (a) $1\frac{4}{11}$ H.P.; (b) $\frac{5}{66}$ H.P.
17. 12,250 ft. lb., 2·97 H.P.   **18.** 1400 ft. lb., 2856 lb. wt.
19. 1344 ft. lb., 16,352 lb. wt.

CHAPTER IX (page 210)

1. (a) 131 m.p.h., N. $15\frac{3}{4}$° W.; (b) 63·5 m.p.h., N. 32° 54' W.
2. 2332 lb. wt.   **3.** 24 lb. wt.   **6.** 84·85 lb. wt.
7. 14·14 ft./sec. due W.   **8.** Upstream at 60° to bank, 8·66 min.
9. 31·6 m.p.h., N. 18° 26' E.   **10.** 300, 400 gm. wt.
11. 13·5 lb. wt.   **12.** 5·77 lb. wt.   **13.** 4·24 oz. wt.
14. 1·732 lb. wt. at 30° with 3 lb. force (between 2 and 3 lb. forces).
16. 1·15 gm. wt.   **17.** 7·07 lb. wt.   **18.** 5·77, 20·82 lb. wt.
19. 115·5, 115·5 lb. wt.   **20.** 546, 386, 283, 546 gm. wt.

CHAPTER X (page 235)

4. 256 ft.   **5.** $\frac{3}{4}$ sec., 24 ft./sec.   **6.** 1 sec.

7. 80 ft./sec.   **8.** 25 sec., 1600 ft., 200 ft./sec.^{2} **9.** 204 ft.

12. 0·0105 sec.

13. 24 ft./sec.: $6\frac{3}{4}$ ft. up, 9 ft. up, and $2\frac{1}{4}$ ft. down from highest point.

14. After 44 sec., 60 m.p.h.

15. (a) 416 ft.; (b) 256 ft.; (c) 165 ft./sec. at an angle of 39° 6′ with the vertical.

16. 126·5 ft./sec.   **17.** (a) After $3\frac{1}{8}$ sec.; (b) $156\frac{1}{4}$ ft.

18. (a) 5 m.p.h. per min.; (b) 2·125 miles.

19. (a) 1500 cm.; (b) 40 sec.

CHAPTER XI (page 252)

5. $1\frac{7}{8}$ lb. wt.   **6.** $\frac{16}{7}$ ft./sec.^{2} **7.** $4\frac{11}{16}$ tons wt.

10. (a) $2\frac{1}{4}$ lb. wt.; (b) $1\frac{3}{4}$ lb. wt.   **11.** 1 : 64.

CHAPTER XII (page 264)

1. (i) (a) 125, 33,600 pdl. sec.; (b) 125,000, 50,400 ft. pdl.

    (ii) (a) $\frac{1}{8}$, 33·6 sec.; (b) 125, 50·4 ft.   **2.** $3\frac{1}{3}$ m.p.h., 172 ft. tndl.

3. (a) 15·7 ft./sec.; (b) 464, 400, 3575 ft. tndl.   **4.** $15\frac{5}{8}$ lb. wt.

5. 23 cm./sec.^{2}, 943 cm./sec.^{2} **7.** (a) 500 kgm. wt.; (b) 479·6 kgm. wt.

8. (a) $1\frac{1}{12}$ ton wt.; (b) $4\frac{1}{8}$ ton wt.   **9.** 7 ft./sec.

10. 175,000 lb. per sq. in.

11. (a) 0·326 ft./sec.^{2}; (b) 97·8 tndl. sec.; (c) 67·2 ft. tons.

12. $13\frac{3}{4}$ tons wt.   **13.** (a) 181·5 ft.; (b) $7\frac{1}{2}$ ft. tons.

14. $6m$ ft. lb.   **15.** $19\frac{3}{11}$ H.P.

16. 4·34 lb. per sq. ft.   **17.** 1875 lb. wt., $468\frac{3}{4}$ ft. lb.   **18.** 1323 ft. lb.

19. 382 metres per sec.   **20.** (a) 5000 ft. lb.; (b) 15,000 ft. lb.; (c) 10 ft.

Index